KELLY

Portrait of a reflective man: Clarence L. "Kelly" Johnson.

KELLY

More Than My Share of It All

Clarence L. "Kelly" Johnson with Maggie Smith
Foreword by Brig. Gen. Leo P. Geary, USAF (Ret.)

SMITHSONIAN INSTITUTION PRESS
Washington, D.C.   London

New in paperback 1989

**Library of Congress Cataloging
in Publication Data**

Johnson, Clarence L.
  More than my share.

  1. Johnson, Clarence L. 2. Aeronautical
engineers—United States—Biography.
3. Aeronautics, Military—Research—United
States. 4. Lockheed airplanes.
I. Smith, Maggie.   II. Title.
UG626.2.J62A36   1985   629.13'0092'4
[B]   84-600316
ISBN 0-87474-564-0
ISBN 0-87474-491-1 (pbk.)

The paper in this book meets the minimum
requirements of the American National
Standard for Permanence of Paper for
Printed Library Materials Z39.48-1984

06 05 04 03 02   10 9 8 7 6

Book design by Christopher Jones

To Nancy

Contents

Foreword

MANY OF YOU, exclusive of true aviation buffs, who pick up this book may wonder who is "Kelly" Johnson? Simply to say that he is one of the most honored and highly successful aeronautical engineers, designers, and builders of aircraft of his or any other time is a fact that is only partially documented by some fifty awards and honors appended to this story. Webster defines genius as "extraordinary intellectual power especially as manifested in creative activity." Though Kelly would deny it, the description fits him to a T.

Aviation, however, is not all or perhaps even the most important element of this story. It is an essential ingredient and backdrop to the unique and insightful story of the man himself that covers a broad spectrum of interest to a wide range of readers. While the story is understated, the reader should be aware of the engineer's penchant for letting the facts, without emphasis or embellishment, speak for themselves.

I first met Kelly in September of 1945, and later had the distinct pleasure, privilege—and education—of working with him and the "Skunk Works" on an almost daily basis for eleven-plus years from 1955 through 1966. This carried us from almost the inception of the U-2, which incidentally was one of the great bargains the American taxpayer ever realized, to the YF-12, the interceptor that should have been built but wasn't, and the first four years of the SR-71, the almost unbelievable "Black Bird"—among other projects. It was a unique and productive experience for me and most regrettably one that may never be repeated for this country. Simply put, Kelly's real legacy is not nearly so much what he has accomplished, but much more how it was done. That is, generally outside—and in many cases in spite of—the so-called regular "system."

The U-2 and SR-71 are two examples of Skunk Works programs that came in on schedule and under contract costs. Still, despite disclaimers, the Skunk Works, Kelly's brainchild—once described by Sen. Sam Nunn as a truly unique national asset and former Deputy Secretary of Defense David Packard as a national treasure—to all intents and purposes has ceased to exist. This is an inexcusable and needless loss for the American taxpayer. Thoughtful readers will question the why of this, as well they should.

This and a great deal more is here in the story of an extraordinary man who certainly has had more than his share of it all.

Leo P. Geary
Brigadier General, USAF (Ret.)
Denver, Colo., 1984

Introduction

CLARENCE L. "KELLY" JOHNSON is the designer of the world's highest-performance aircraft—the big bold "Blackbirds," the SR-71 and YF-12—that were flying secretly at three times the speed of sound while other experts still were insisting that it was not feasible; and the graceful, glider-like U-2, which can attain altitudes admitted to be "above 80,000 feet."

He designed America's first operational jet fighter, the F-80 Shooting Star. His dramatic twin-boomed P-38 Lightning fighter-interceptor of World War II was the first aircraft to encounter the phenomenon of "compressibility" as the wing's leading edge built up supersonic air turbulence. He has contributed to the design of more than 40 aircraft, more than half being his original design.

He holds every aircraft design award in the industry, some for the second and third time: the National Medal of Science; the National Security Medal; and the Medal of Freedom, the highest civil honor the President of the United States can bestow.

His "Skunk Works" at Lockheed—more formally, Advanced Development Projects—is recognized worldwide as unique in its record for turning out "breakthrough" designs in minimal time and with maximum security. "Be quick, be quiet, be on time," are Kelly's watchwords.

When the Russians in 1960 exhibited to the public in Red Square the wreckage of the aircraft they billed as the U-2 in which they had downed Francis Gary Powers, Kelly's response

With wife Nancy Johnson, during 1983's presentation by President Ronald Reagan of the National Security Medal.

to press query when shown the photo was typically direct and dramatic: "Hell, no," the aircraft designer barked. "That's no U-2."

The Russians had downed the U-2, untouchable for years at its high altitude on reconnaissance flights over Russian territory; but Kelly blew their act. A designer who went into the factory and participated in every phase of design and development as well as production, he recognized immediately that the mangled parts the Russians had displayed were not from any U-2.

Controversy is nothing new to this much honored engineer.

In his first day as a just-graduated engineer on the job that was to last 44 years, he told his employers that the new, all-metal aircraft with which they planned to challenge the air transport field was unstable! While such instability was commonly accepted in aircraft of the 1930s, the young engineer stubbornly refused to accede to the view of the professors with whom he had performed wind tunnel tests on the model at the University of Michigan. They were willing to accept the imperfections. Kelly was not. He was right, of course. The reworked result was the first in the long line of twin-tailed Lockheed transports that would make the company's name known around the world in the 1930s and '40s. This characteristic behavior soon earned him the nickname, "the Old Goat," among the Lockheed engineering staff.

Kelly always held to his principles.

He advised the U.S. Navy in the early '50s that a vertically-rising aircraft for which his company had a development contract was unsafe, with the limited engine power then available, and should be abandoned.

He refused to go ahead with a hydrogen-powered aircraft—ahead of its time in the late '50s—and he turned back a development contract after initial work indicated the plane would be a "wide-bodied dog," in the words of his successor at ADP.

He returned to the U.S. government approximately $2 million saved on the $20 million U-2 contract, having produced an extra six aircraft for the same money intended to cover 20 aircraft.

"I have known what I wanted to do ever since I was 12," Johnson says.

Now officially retired, he continues in an advisory capacity at Lockheed, where he maintains an office in the "Skunk Works." He still works to a busy schedule, though not always now beginning at his once customary 6 a.m., the better to communicate with East Coast military offices operating with a three-hour headstart.

This book is not intended to be a history of aviation nor a documentation of specific aircraft development. It is the personal reflection of one man in his time.

Maggie Smith
Sherman Oaks, Calif.

1

Poor But Not in Spirit

NORTHERN MICHIGAN IN MID-WINTER is harsh, cold country to a young immigrant seeking to carve out a new life. My father didn't choose it intentionally.

In the year 1882, at age 24, he left the small city of Malmo in his native Sweden for a better life in the United States, leaving behind his intended bride, Christine Anderson. Sweden had universal military service. Peter Johnson was about to be conscripted into the army, and he did not want to carry a gun.

He had saved $600 with which he intended to buy a farm in Nebraska. His future would be in the fertile plains of the Midwest.

He got as far as Chicago before he found that all was not opportunity in the "new world." There Peter fell in with bad company, some fellows ready to take advantage of an unworldly foreigner literally just off the boat. He paid them his $600, thinking he had bought the farm he wanted in Nebraska. His new friends put him on a train headed, instead, for upper Michigan. He got off at Marquette. It was winter and awfully cold, he thought, for Nebraska.

When the truth became obvious, he was faced with the need to support himself in a strange country. He was a mason by trade, but the only work available there in midwinter was on the railroad, laying ties. That's what he did for a time until he could make connections with a local construction company and work again as a bricklayer.

It took several years before Peter could afford to send for

Christine. They were married and settled in the small mining town of Ishpeming, not far from Marquette. That is where I was born, on February 27, 1910.

I, Clarence Leonard Johnson, was the seventh in a family of nine children. We were very poor, and all of us learned early in life that as soon as we were able we had to help earn what we needed.

My earliest memories of my birthplace are of how beautiful it was. Even the long railroad trains, their cars piled with iron ore, going off both east and west seemed beautiful to me as I watched them from the bluffs above town. I'd go out into the woods, summer and winter, and always had a little hideaway camp where I could go with my dog, Putsie.

We lived in a succession of three large houses, all rented. I remember best the last of these, a big frame house painted green, on top of a hill, on Summit Street. It was bitter cold in the winter. To fuel our wood-burning stove, my father would pick a bright clear day to hitch our horse, Mac, to the cutter and drive four or five miles into the woods to cut birch. When I was about eight years old, I began to join him regularly. It was so cold that my mother put a jug of hot water by my feet, then wrapped me up to keep me warm. She packed our lunch in a dinner pail, too—some hot coffee and sandwiches. We would look for a fallen tree since that was easiest to cut up. We had a crosscut saw, and I would try to help by pulling one end. But my father generally did a lot of the cutting with an ax. Our day's work done, we'd pile our wood in the back of the cutter—at least three wheelbarrows-full—and head home with a comfortable feeling of accomplishment. Our Christmas tree each year came from the woods, too. I never recall seeing anyone else on these forays. The area was not very widely settled.

We had both wood and coal-burning stoves for cooking and heating in that house, and we got our coal in another thrifty manner, picking it up from the railroad tracks. The train always dropped some coal, and one gunnysack-full was enough for our needs for a day. My sister Alice and I would take our sleds—we each had one—and gunnysacks and fill them up after school. The engineers came to know us; and if there wasn't a supply of coal that had fallen off the trains, they would throw some off.

The severe winter weather made it difficult for my father to work. Snow and ice would form on the bricks so that they could not be joined. But on relatively warm days, he would solve that problem with an old, empty 50-gallon oil drum. He knocked holes in the bottom to get a draft and built the biggest fire he could inside, having placed the frozen bricks around the out-side. Then, before ice could refreeze on them, he would lay the bricks as quickly as possible. His work always was subject to weather; but when he worked, my father could lay 2,000 bricks in a day.

My mother worked, too, in addition to raising a family and running the house. She took in washing, from the wealthier people in the town. Each basket of clothes brought in a couple of dollars. Every day she washed, and not in a washing machine but on a scrub board. Our basement wasn't large enough to hold a wash; so every day, summer and winter, she would hang it outside.

In winter, of course, it froze and had to be ironed dry. The older girls, Ida, Freda, and Agnes, helped with this work. I did, too. I picked up and delivered laundry at least a couple of times each week in my wagon, or on my sled in winter. I did not like to be seen doing this, and I remember particularly one time when there was a Saint George Day parade and the main streets were full of people. I took back alleys all the way home. I vowed then that one day I'd return to Ishpeming and not on the back streets but the best streets.

I loved the woods surrounding Ishpeming and always had a secret hideaway there in a place southeast of town that I called Surprise Valley, because it was pretty well hidden from view. I've gone out in winter when it was ten degrees above zero. I wouldn't build a fire but would listen to the trees cracking in the cold, try to read fox and other animal tracks in the snow, and chase a few rabbits. I learned how to build a lean-to from reading a book on Indian explorers. I'd cut pine boughs, run a sturdy limb two or three inches in diameter between two adjacent trees, then weave the boughs together stems up, so the branches would fall down like shingles. It made a shelter that was quite waterproof. The same technique of weaving boughs made a floor. I'd leave the front open so I could watch the animals. And I usually had a little food stored there—mostly just some bread and butter in a tin can. When Putsie was with me, we'd share it. I'd make a new lean-to every winter, because in summer it would be torn down or dried out. It took me only about a day to build one with a lath hatchet borrowed from my brother Emil.

School always was interesting to me. I was eager to go and would show up early to be first in line to enter the building. But there was one fellow who took to pushing me out of the way when he arrived later. And he made the mistake of calling me Clara for Clarence. His name was Cecil. Unfortunately for me, he was about a foot taller than I was, a long gawky kid; so when I realized that action had to be taken, I knew it would take some planning.

The occasion came during one recess in the schoolyard when words came finally to blows. The other kids pushed us together, and I knew there was only one way I could lick this guy. So I kicked him behind the knee to trip him, then jumped on him when he fell. There was a loud popping noise. I had broken his leg.

The principal, Mrs. Lacey, and my second grade teacher, Miss Hass, didn't quite know what to do. There was Cecil with a broken leg, and I not only admitted doing it but on purpose. They spanked me over the knuckles so hard with a ruler that it broke. But I didn't cry and that brought me favor with the other children.

Cecil was from one of the wealthier families in town and I expected his family to raise trouble with ours. I was somewhat afraid to go home. It was the only time I ever feared a whipping. Despite my mother's assurances when I got there that she was not going to spank me, I announced that I wasn't coming in, and I ran off to my hidden camp in the woods, where I spent the night. It was late spring; so it wasn't cold. Putsie and I shared the store of old bread and butter I had there. I returned about three o'clock the next afternoon, having missed school, but warmly welcomed at home.

When I returned to school, the kids had decided that I didn't act like a Clarence and should have a good fighting Irish name. There was a song popular at that time, "Kelly from the Emerald Isle," and they sang a verse about "Kelly with the green necktie." They named me Kelly. It stuck.

We were really hard-pressed for money in the early years, and my mother not only took in washing but scrubbed floors. One day she got a job to scrub the floor of one of the big stores in town, Sellwood's. It looked to me like almost an acre of floor area. But Alice and I joined my mother at the job and in one day we had the oak floors scrubbed clean.

We were so poor at that time that when we would take a can to buy kerosene for our lamps—for 15 or 20 cents—we would plug the spout with a potato, then when we had carried the kerosene home, cut away any part of the potato that had been sloshed with kerosene and save the rest to be eaten.

To help with finances, I spent one summer with an aunt in the farming community of Sands, less than 30 miles east and slightly south of Ishpeming. I earned my keep there, doing things like raising the gear ratio to 20-to-one on the cream separator to make that a lot easier to operate. I also earned $31 picking wild blueberries. I got $1.00 a peck, which took a whole day to collect. When I went home at the end of the season, I handed the entire $31 to my mother. There were tears in her eyes as she thanked me, she was so touched that I had not kept anything at all for myself. No contribution I have ever made since has made me feel happier; none has been more important to me.

The next summer Alice and I decided we both would like to make some more money that way. She and I were only three years apart in age and she was my closest companion in the family. Clifford and Helen, seven and eight years younger, were too young to keep up with us.

Alice and I packed our things in a single suitcase and headed for Sands on the train. It was a six or seven mile walk from the railroad station to our aunt's house, on a hot June day, and when we got there we were told that we could not stay that summer. So, we picked up our suitcase, put a long branch through the handle so we each could carry one end, and trudged back to the train station for the return trip home.

Our parents were stern but not severe, serious in manner but considerate of us children. I never was struck by either of them. We children were expected to show responsibility. I had access to my father's tools and a workshop from the time I was seven or eight years old. I could use any of his tools I could handle, so long as I didn't break or lose them and always put them back in place.

Some of my earliest lessons in construction were in watching my father build toys for me. One winter day, in a very, very cold barn which housed his shop, he constructed a rocking horse from a piece of birchwood, half a dozen pieces of wood, and some rope. It was a fine, sturdy horse when finished—painted white, with the birch body section left in its natural state.

He also built me a wagon, painted green, with a remote control brake that I could operate with either knee or hand. The wheels were purchased, as was some hardware, but the rest was the product of his workmanship. It, too, was very sturdily constructed, and I used that wagon for many years.

My father was very mechanically inclined, too, as well as being a proficient carpenter and mason. An all-around craftsman. My respect for tools and machinery I learned early in life from his example. He worked throughout his life in the building business and taught me a great deal about construction. It was to be very useful to me.

My father's hands were so gnarled and calloused from handling rough bricks under every sort of condition that he ceased to have much feeling in them. The hard life once caused even him to rebel. He spent one whole paycheck on drink. It was the only time.

Our parents instilled a love of learning in us children. They always encouraged us to study in school and to read on our own. Next to my father, I credit Andrew Carnegie with being the most important influence on my early life through the library he had donated to Ishpeming—as he had in many other small towns whose natural resources had helped build his fortune. In Ishpeming, of course, it was the iron ore. He returned an even richer resource.

I went to the library almost every day with Putsie. It opened a whole new world to me. I discovered Tom Swift and read not once but several times *Tom Swift and His Aeroplane, Tom Swift and His Electric Automobile, Tom Swift and His Submarine,* and on through the entire series. Tom Swift was a very highly skilled designer, engineer, pilot, and operator of many kinds of locomotion, and an adventurous young man. It became my goal to be just like Tom.

I read other books on aircraft—the Rover Boys, Collins' book on model airplanes—and decided by the time I was 12 years old that I would be an aircraft designer. My whole life from that time was aimed at preparing for that goal. I put together my first book on aircraft, mostly from clippings, and designed my first plane—the Merlin battleplane—named for the magician of King Arthur's court. I made hundreds of model airplanes.

The other kids resented my studious habits as well as the good marks I consistently got in school and would taunt me. I didn't let it bother me. But it made me one of the fastest runners in Ishpeming. The kids would lie in wait for my daily return from the library and in winter would fire off snowballs stuffed with coal chunks.

There was a fearful period of two weeks, though, when my future was in serious doubt when I was hit in the left eye with an arrow fired by Helen while playing cowboys and Indians with her and Clifford. I was blinded.

My mother's training and experience as a nurse helped her keep calm and take charge. She had served in the local hospital during the dreadful influenza epidemic that spread worldwide after World War I. None of our family contracted the disease, and Mother worked long hours at the hospital while Agnes and Alice took care of the household.

She carefully removed the arrow, which had entered alongside the eyeball, mopped up the blood, and determined that the eyeball itself had not been hit. But I was in such a state of shock that I lost sight in both eyes. I was terrified. It was two long unforgettable weeks before my sight returned.

I continued to read everything I could get my hands on about aviation. The war had made the airplane a reality, and every now and then we actually could see one as a barnstormer came through selling rides for $3 apiece.

My enthusiasm for the subject spilled over into my more conventional studies. One day while I was expounding on current events—aviation, of course—the school principal, then Mr. Walter Griese, visited our classroom. He thought it was good that the kids were studying such subjects outside the routine curriculum, and he invited me to address a luncheon meeting of the local Lions Club on the future of aviation.

This was quite an honor for me, and my parents were so proud that they bought me my first pair of long pants for the event. In white shirt and tie, I made my debut as a speaker on aviation. I was still so short that they stood me on a chair to speak. I was pleased to receive quite a lot of applause.

When I was about ten years old, I saddled our horse, Mac, and rode out to western Ishpeming where Emil was working at lathing, and I learned that trade. He and my older brother Arthur already had married and moved on, as had the older girls, while I was growing up. There was quite a few years' difference in our ages, and I hardly knew them.

The building business was picking up in Ishpeming, and at an early age I could work as a lather. Laths are wooden slats, four feet long, that are nailed on studding and two-by-fours in new construction as the base for the finished coat of plaster or wood siding. I became quite proficient. By the time I was 12 years old, I was contributing $7 every week for room and board. I had to make $10 to have a little left over for myself. I could put

An impoverished but high-spirited and determined youngster, Kelly with his younger brother Clifford. Below, a book written at age 12, an example of Kelly's early interest in the fascinating new field of aviation.

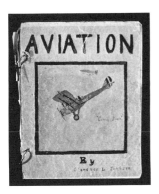

on 40 bunches of laths a day at 25¢ per bunch. From then on I was self-supporting.

The improvement in the building industry convinced my father that he should move our family from the small city of Ishpeming to Flint, a much larger city offering more opportunity. So in 1923, when I was 13, we moved to Flint, some 300 miles away on Michigan's lower peninsula.

We couldn't take a dog with us in the passenger car of the train, and we could not afford the cost of shipping by crate. My last view of Ishpeming is of Putsie running along beside and then behind the train, trying to keep up.

2

A Good Move

THE FORTUNES OF OUR FAMILY improved significantly with the move to Flint. There was lots of building, and my father was able to establish himself in the construction business. I worked for him and others all through high school and junior college. And my mother no longer needed to take in washing nor work outside the home.

My goal remained to become like Tom Swift, and I studied toward that end. Flint had an excellent public school system, and, as before, I enjoyed very much attending classes. The city had an even larger library than Ishpeming's, and I quickly became a regular visitor there, too.

It was a year yet before Charles Lindbergh's historic transatlantic solo crossing would awaken the world to the excitement and potential in the air, but in 1926 there was enough stirring of interest that the city's Kiwanis Club sponsored a model airplane building contest for schoolchildren.

I carved out a model of my Merlin battleplane, and it won second prize—$25. That didn't cause much excitement at school. I was always talking aviation, making it the subject of my reports when given a choice, and getting A's in my studies.

After graduation from high school, I was tempted to digress from my dedicated course. After all, I had worked hard and saved $350 from lathing, and I thought I owed it to myself to take a ship and work my way around the world, have some adventure. But one of the dedicated teachers it has been my good fortune to know, Miss Bertha Baker, spent most of one afternoon explaining to me why there should be no gap between high school and junior college. She convinced me. And I entered Flint Junior College.

In junior college, I was able to take engineering courses for the first time. I studied physics, mathematics, and calculus. I reached the point where I could tutor in calculus and make some money. I loved mathematics and still do. It was a very good junior college, and I received a solid background for my more advanced university courses later.

As often as I could on vacations and weekends, I worked at lathing. My average pay per day was $10 to $12. I'd have to put up 2,000 laths for that $10. That took a lot of nail pounding.

In summers, I also worked for the Buick Motor Car Company, swinging fenders on the production line or working on motor repair and block test. I would be so dirty and oily after a day's work that I was not allowed to sit down on the streetcar going home. It really didn't bother me too much. I was reading and studying all the time, standing up. In that period, I even tried to understand Einstein. Only 12 people in the whole world were supposed to be able to do so; I wanted to be the 13th!

At long last in Flint I had my first airplane flight. It cost $5 for three minutes. It was a big clumsy machine—an old Standard biplane, a four-passenger model with a single pilot. We got all the way up to 700 feet before the engine quit and we had to make a forced landing. But it was fun! It was noisy, it was drafty, it was great! I still wanted to design airplanes.

Actually, I began to think maybe I should learn to fly because all of the early great designers—Glenn Curtiss, the Wright brothers, Glenn Martin—they all knew how to fly.

So one rainy morning after I had been graduated from junior college but had not yet enrolled in a university, I went out to Bishop Airport in Flint, prepared to hand over my entire fortune at the time, $300, for 10 hours of flight instruction. In the shack that served as his office I found the pilot, Jim White. He talked to me for some time, inquiring about what I hoped to do with my future.

"Kelly," he said then, "you don't want to start off on your career by giving me $300 to learn to fly. That won't get you far enough. You have good grades; you will go a lot farther if you go on to the university. I won't take your money. You don't want to end up as an airport bum like me."

Well, he needed that money perhaps even more than I did. But he was a big man. It was another fortunate encounter for me to meet another person who guided me with wisdom at a critical point in my life. I took his advice, and I've always been thankful for it.

Flint Junior College had a championship football team. I played on it. After graduation, several of us were offered scholarships at a university in the South known for its athletic prowess. I went down during summer vacation to get in some football practice.

When it came time to enroll, I was planning to select courses that could lead to my transferring to aeronautical engineering in another university after a year. I soon found out what my options were.

"Here, kid, here's your curriculum," the coach told me one day.

"But I haven't chosen my courses yet," I said, surprised.

"Yes, you have," he insisted. "You're a coach's assistant. You're taking physical education."

"But, sir, I'm going to be an engineer. I've wanted to be an airplane designer all my life. I want to study aeronautical engineering," I protested.

"Kid, you're a coach's assistant." He repeated, "You're a coach's assistant. Take it or leave it."

"Not me." And that was that.

My next move was to phone the University of Michigan about athletic scholarships. They offered them, and my grades were good enough for admission, so I got in my trusty Ford Model T roadster and drove up to Ann Arbor to try for a scholarship there. My $300 would do no more than pay the tuition.

Just about the second thing I found out was that an undergraduate was not allowed to have a car on campus. So I decided to take my car home and then come back to try out for football.

On the way, I was forced off the road and across a culvert by a big Pontiac. I was draped across the windshield and had a deep gash in my forehead. Worse than that, the cut later became infected. I couldn't go out for football.

It was one of the best things that ever happened to me, because now I had to try to find work in an engineering line. I did, but not before I had washed at least 10,000 dishes, as many glasses, even more silverware, and carried out tons of garbage working in a fraternity house that first semester. The year was 1929, and there was almost no building at Ann Arbor so I couldn't work at lathing.

The best thing about that kitchen job was the way we were treated by the wonderful black cook. She saw to it that her twelve fellows in the kitchen working for their meals were fed first before the fraternity brothers—and with the best portions.

After one semester, I became assistant to Professor Edward A. Stalker, head of the aeronautical engineering department at the university. It was a job I was to keep throughout my university career. But more importantly, it was my first work in engineering.

3

Becoming an Engineer

THE UNIVERSITY OF MICHIGAN dates from 1817 as a territorially-chartered college in the then-frontier town of Detroit. The present campus was established at Ann Arbor in 1837. It is one of the early great universities, and a beautiful one—its big brick buildings of classical design, ivy-covered, amid neat lawns bordered by trees that flower in season, on a spacious, sprawling campus. It can be at once formidable and welcoming to an incoming student.

But the real beauty of it for me when I enrolled in 1929 was the distinguished faculty. Many had impressive national, even international, reputations in their fields. I thought I never could be so smart as these men. I couldn't wait to begin my classes.

At that time, in order to get a degree in aeronautical engineering you had to study all the different fields of engineering—civil, chemical, electrical, mechanical—leading to the study of aeronautical engineering. It was an excellent curriculum because it provided a very good basic education in everything it took to design and build an airplane.

My first professor was Felix Pawlowski, teacher, pioneer aircraft designer, and philosopher. The Polish and Russian universities had been into aviation early and were in the forefront of aeronautical knowledge at that time. Pawlowski, a Pole, had worked in Russia with Igor Sikorsky on the world's first four-engine airplane in 1913. He had been trained by Professor Alexandre Gustave Eiffel, designer of the Paris tower, and worked with him on a wind tunnel. Professor Pawlowski was responsible for bringing the first wind tunnel and the first aeronautical engineering curriculum to the University of Michigan.

He taught my first course in aerodynamics and helped me get the first engineering jobs that would pay my way through school. He, like some of the other professors, had contracts outside the college. In the wind tunnel, I worked for him on design of the Union Pacific streamlined train, on a smoke-removal project for the city of Chicago, and on one of the very early proposals for generating energy with a wind machine.

The professors were broadminded people, with interests and contacts outside the university. They took a personal as well as professional interest in their students. One day Professor Pawlowski taught me an important lesson in keeping an open mind. He took me down to a bank vault where he had some wax impressions of hands, "spirit" hands, he had from a seance. They were entwined in a manner that could not be explained. This eminent scientist was willing to consider their validity. He wanted me to learn to keep an open mind.

"Don't automatically write anything off," he said. "Anything." I've remembered that.

Professor Edward A. Stalker, author of a basic text on aerodynamics and an outstanding mathematician, was head of the aeronautical department. He, himself, not some registrar, helped me plan my course of study.

After I had shown enough academic progress he selected me as student assistant and I was able to earn enough money to quit the lowly job in the fraternity kitchen.

As head of the aeronautical engineering department, Professor Stalker operated the wind tunnel and got me involved in wind-tunnel testing.

One day I asked Professor Stalker, "Could I rent the wind tunnel when it's not needed by the university and get some jobs on my own?"

"Sure," he said.

So for $35 a day, plus the power charges, my best friend in college, Don Palmer, and I became part-time proprietors of the

University of Michigan wind tunnel. The money didn't mean anything to the university; renting the tunnel afforded them a chance to see what the students could do.

Immediately, I approached the Studebaker Motor Company. It was obvious that the wind tunnel could be very useful in designing streamlined automobiles. We got an assignment to test the Pierce Silver Arrow, which was to become one of the early "totally-streamlined" cars. We knew all the tricks on how to reduce drag caused by air resistance. We found, for instance, that the big ugly headlamps on Studebaker cars were eating up 16 percent of the power the engine developed at 65 miles an hour. We managed to get them shaped into the fenders. We worked on a lot of other problems and ran many, many tests.

So we worked not only for Professors Pawlowski and Stalker but for ourselves. Tutoring in calculus also brought me $7.50 an hour.

Some of the courses, seemingly not much related to aircraft, turned out to be especially useful to me later.

Mechanical engineering, for example. For our final examination, we students had to make a heat balance evaluation of the university's power plant, a large steam facility with four big boilers, that provided not only heat but power. I, an aeronautical engineer, was put in charge of the other mechanical engineers for the three-day continuous test, measuring coal input and then accounting for all the energy through the entire process down to the ashes. It was a valuable lesson in energy balance.

Professor O. W. Boston was a pioneer in the field of machining metals, and the author of a book on the subject. He also worked with the automobile industry on methods of cutting such materials as high-strength steel and other metals to improve machinability and thereby improve efficiency in automobile production.

He was, I believe, the first to imbed thermocouples in a tool on a lathe or milling machine to measure tool temperatures. He also worked on the design of tools to remove metal rapidly. Much later I was to apply his methods in the approach to machining new metals, specifically, titanium and stainless steel.

The two Timoshenko brothers, Russians, were professors of structure. The study of vibration and structure was very important, basic for me, in learning how to make aircraft wings and tails that wouldn't flutter. That was a bugabear of a problem in those early days.

There was Professor Milton Thompson in aeronautics, Professor Walter E. Lay for engines, others whose names escape me now, each an expert. This was heady stuff for the kid from Ishpeming who wanted to be like Tom Swift. It was an exciting adventure, associating with some of the best minds in every field important to a would-be engineer.

Respectful though I was of the great experience and knowledge of my professors, I yet was not so deferential that I would not argue back if I disagreed. And I did.

With Professor Pawlowski, who had given me a B grade on my computation of wind-tunnel tests on a little biplane—I didn't like to get less than an A— I argued that my numbers were correct. I proved it, and he changed the grade to A. He kept an open mind himself, as he advised.

When Professor Stalker published a new text, I had the temerity to write out all the answers to all the problems and proposed to publish them. I was persuaded not to do this since it would undermine the book and considerably diminish sales.

These men were not only my mentors but my friends and companions, and not because their losses at poker gave me another small added income.

But most of the time I was working or studying, correcting papers or tutoring. I completed three years' university work in two. There was little time to play. I knew I had to work hard to become a good engineer, and I enjoyed it.

There were lighter moments. Like the time Don and I were cleaning the wind tunnel and became so high on gasoline fumes and so noisy that we disrupted nearby classes.

And lonely times. Trudging back across campus after dinner—which cost perhaps as little as fifty cents or as much as

$1.25—for more wind tunnel work or correcting papers for the professors.

Those were the days of Prohibition and home brew. I was very much against drinking then, believing the stories of my hard-drinking Swedish ancestry, but my colleagues in the boardinghouse imbibed the stuff they brewed—fermenting fruit juice in a can. Returning from campus on a winter evening, I found one fellow sitting outside in the snow wearing only his shorts, and mumbling to himself. Another convincing argument against drinking.

I had my first ulcer in high school, but in college I had one all the time. I've always been a worry wart. I discovered that if I kept something in my stomach all the time I felt fine. I did that with repeated ingestion of two doughnuts and a glass of milk—at a cost of twenty cents. One semester I computed that I had consumed 647 doughnuts—at five cents each.

In all my time at the university, I went out on dates just twice. Once to a good movie—I don't remember what it was. And to a class dance. Dancing came easily to me, fortunately. When would I have had time to practice? I had learned in high school. But I didn't have time for romance either, and I deliberately avoided any entanglement. There would be no detours from my goal.

The year 1932, when I was graduated with a bachelor's degree in aeronautical engineering, was not a propitious year for job hunting. My friend Don and I investigated opportunities on the East Coast—Sikorsky, Martin, Curtiss—with no encouragement. We decided to join the Army Air Corps and become aviation cadets. We would learn to fly and test airplanes and learn all about them. I had passed every other entrance requirement and then took the eye examination. My left eye, the one nearly lost to the arrow in a childhood game of cowboys and Indians, was not up to the standard required by the Air Corps although it never had given me any trouble. Once again, an accident changed the course of my career. If I'd been accepted as an aviation cadet, I would have taken that route and stayed with it.

Don and I then borrowed Professor Walter Burke's Chevrolet and set off to tour the aircraft plants in the West looking for work as engineers.

We were short of money, of course. Our earnings in the tunnel and elsewhere had gone toward school expenses. To get better mileage from the professor's car and save money on fuel, we drilled a one-eighth-inch diameter hole in the manifold inlet and inserted a valve, so while driving we could open it and lean out the fuel in flight, so to speak. We did get three or four more miles to the gallon.

To hold down expenses, we'd buy milk, bread, and sandwich meat to make our lunches. We camped out in schoolyards, by the side of streams, in fields, wherever we could, and nearly ended our careers early when we picked a site in the dark one night and were awakened by a passing railroad train uncomfortably close. We decamped hurriedly.

That was the most excitement we encountered until we got to Lockheed in Burbank in the San Fernando Valley of California. The company had been purchased from receivers by a small group of aviation enthusiasts just that June for $40,000.

The company was in the process of being reorganized. Lockheed already was a big name as designer and builder of fast plywood aircraft flown by many of the famous names in early aviation. There were no jobs for engineers yet, but Richard von Hake, chief engineer at Lockheed when it was part of Detroit Aircraft and who was to become production manager of the new company, suggested:

"Look, something is going to come of this. Why don't you go back to school and come out again next year. I think we'll have something for you."

Well, there were no other job opportunities that year; we'd tried all the principal companies.

So, we returned to the University of Michigan for a year of graduate study. To afford this extra year to get my master's, I applied for and was awarded the Sheehan Fellowship; the $500 paid my expenses. I majored in supercharging of engines, to get high power at high altitude, and boundary layer control, how to control airflow around fuselage, wings, and tail. It later proved a fortunate choice. Of course, aerodynamics is basically boundary layer control. And I always loved engines and aerodynamics. It was a natural choice, as well.

In our graduate year, Don and I did much more wind-tunnel testing on our own as well as for the university.

The local newspaper reported, "Five of the qualifying cars which will race at Indianapolis Memorial Day have bodies designed by two University graduate students, C. L. Johnson and E. D. Palmer. All of the cars are semi-stock Studebakers and all qualified for the race at speeds ranging between 110 and 116 miles an hour. . . ."

We managed to improve the miles per gallon on these cars from seven to 11.6 at 113 miles an hour. That was important, because in those days fuel-tank capacity was limited.

A few proposals we explored, such as streamlining wheels, the drivers refused to accept. I was given a demonstration of the argument against that one day when we were at the track for tests.

It was very exciting circling the track at speeds of 130 to 140 miles an hour; but if you had solid-disc streamlined wheels, the wind across the track would just pick up the car and set it down again about four feet off course.

Another idea I tried to sell was dive brakes on the side of the car, because the streamlined cars would reach such high speeds on the straightaway that they would lose all their advantage by having to brake at the turn. This was a difficult effect, too, because if one brake opened a bit earlier than the other, the car would just swap ends. They still don't use dive brakes today to my knowledge; but many other ideas from aerodynamics have been incorporated in the design of racing autos.

The experience was a liberal education for me in the practical application of aerodynamic theory.

Among the airplane models tested by the university was a new design from the recently-reorganized Lockheed company. The chairman of the board, Robert Ellsworth Gross, 35, had decided that the company's future was not in the single-engine wooden aircraft that had been so successful in the past, but in the newer all-metal designs with twin engines and the capability of carrying more passengers.

The new model was the Electra. It developed some very serious problems, I thought, from what I then knew of aerodynamics. It had very bad longitudinal stability and directional-control problems. But most aircraft of that day had similar failings. Professor Stalker, in consultation with Lloyd Stearman, already a recognized top-notch designer at age 33 and first president of the company, decided the figures were acceptable.

When I left college with my master of science degree in 1933, I owed only $500 and had enough money to buy a used Chevrolet sedan to try again for a job in California. Don went with me, and once again we modified our car to stretch gasoline. Thanks to our work in the wind tunnel and the Sheehan scholarship that last year, I was relatively wealthy. We didn't try to continue our consultancy work with the wind tunnel because it now was so lucrative that it was attracting the professors' attention. Besides, at the university we certainly weren't going to design aircraft—and that was my goal. But I didn't make that much money again until 10 years later.

When we got to California in 1933, I was hired at Lockheed by Cyril Chappellet, one of the original investors and now secretary of the company, assistant to the president, and personnel officer, and by Hall L. Hibbard, chief engineer. Both were young men themselves. I think an important reason for my being hired was that I had run the wind-tunnel tests on the company's new plane. I was to receive $83 a month to start in tool design until they could assign me as an engineer. There were five engineers at the time, counting Hibbard. Don Palmer was hired at Vultee Airplane Company in Glendale.

Practically the first thing I told Chappellet and Hibbard was that their plane was unstable and that I did not agree with the university's wind-tunnel report.

4

A Growing Airplane Company

WHEN I ANNOUNCED AT LOCKHEED that the new airplane, the first designed by the reorganized company and the one on which its hopes for the future were based, was not a good design, actually was unstable, Chappellet and Hibbard were somewhat shaken. It's not the conventional way for a young engineer to begin employment. It was, in fact, very presumptuous of me to criticize my professors and experienced designers.

Hibbard didn't comment on that first day, but he thought about what I had said.

He grilled me thoroughly on my background. Could I draw? How much math had I had? Well, I'd had quantum theory and had tutored in calculus. I had good grades, recommendations from my professors, and the wind-tunnel experience.

Hibbard himself was a fine engineer, with a degree in aeronautics from Massachusetts Institute of Technology, an outstanding institution. And he wanted to get some "new young blood . . . fresh out of school with newer ideas" in the engineering department, as he explained many years later in an interview.

"He looked so young," he said. "I was almost afraid that he couldn't read or write! . . . We got some fresh ideas, believe me! When he told me that the new airplane we had just sent in (to the university wind tunnel) was no good, and it was unstable in all directions, I was a little bit taken aback. And I wasn't so sure that we ought to hire the guy. But then I thought better of it. After all, he came from a good school and seemed to be intelligent. So, I thought, let's take a chance. . . ."

As a start, I was assigned to work with Bill Mylan in the tooling department, designing tools for assembly of the Electra, until there was more space in the engineering department and a job for me there. Mylan was an old hand and knew his business.

"I'll build them, kid, and you can draw them later," he explained to me.

I learned some useful lessons. For one, I learned to read the fine print. One of my first jobs for Mylan was to design a heat-treat furnace for the new dural aluminum materials that would be used in production. I didn't know much about such furnaces, so I went downtown where there were several in operation and studied them, then went back and drew up what I thought we would need. A few days later, I went out in the shop to see how it was coming. The workmen were standing up to their ankles in brick chips. They had a big, powerful bandsaw and were trimming bricks.

"What are you doing? Just lay the bricks in there. Why are you cutting them?" I wanted to know.

"Mr. Johnson, we're just following your orders."

In the corner of the shop order was a line in small print, "Unless otherwise specified, all tolerances are to be met within $\frac{1}{32}$ of an inch." So they were sawing the bricks because I had listed 2½ by four by nine inch brick.

I discovered that there was a lot to learn about tooling. My first design for a jig—that's a pattern or framework in which the airplane and its parts are built—allowed room to work on one side only—unless the workmen crawled under it and worked over their heads!

After a few months, Hibbard called me into his office.

"Kelly, you've criticized this wind-tunnel report on the Electra signed by two very knowledgeable people. Why don't you go back and see if you can do any better with the airplane?"

Hibbard sent me back to the University of Michigan wind

tunnel with the Electra model in the back of my car. It took 72 tunnel runs before I found the answer to the problem.

It was a process of evolution. On the seventy-second test, I came up with the idea of putting centrollable plates on the horizontal tail to increase its effectiveness and get more directional stability. That worked very well, particularly when we removed the wing fillets, or fairings onto the fuselage—put on apparently because they were coming into style and being used successfully on such airplanes as the Douglas DC-1. And we avoided the trouble others had with them when not used properly.

We then added a double vertical tail because the single rudder did not provide enough control if one engine went out. That was so effective we removed the main center tail. And there you had the final design of the Electra. The distinctive twin tails on all of the early Lockheed metal airplanes, and the triple tail of the familiar Constellation airliner of the mid-'40s and '50s, were the result of these tunnel tests.

I have saved a letter Hibbard sent me while I was still working at the university wind tunnel. He had airmailed some cowls for a more powerful engine, with 550 horsepower, that Pan American Airlines wanted in the airplane. The airline was "very, very much interested," he wrote. ". . . in fact, want them so badly that they are actually going to pay for these last tests which you are running up there now."

"Dear Johnson," the letter began. "You will have to excuse the typing as I am writing here at the factory tonite and this typewriter certainly is not much good.

"You may be sure that there was a big celebration around these parts when we got your wires telling about the new find and how simple the solution really was. It is apparently a rather important discovery and I think it is a fine thing that you should be the one to find out the secret. . . . Needless to say, the addition of these parts to the horizontal surfaces is a very easy matter; and I think that we shall wait until you get back perhaps before we do much along that line."

Some specifics followed on the cowl tests for Pan Am. Then,

"Well, I guess I'll quit now. You will be quite surprised at the Electra when you get here, I think. It is coming along quite well. Sincerely, Hibbard."

It was typically generous of him to stay at night and type a letter himself in appreciation of the work of a new, young engineer. It meant a great deal to me.

When I returned to the plant, I was a full-fledged member of the engineering department. I was number six. There were James "Jimmy" Gerschler, George Prudden, Carl Beed, and Truman A. "Tap" Parker. The quarters weren't much, the roof leaked, but I was an honest-to-God aircraft engineer. I worked not only on the aerodynamics of the airplane, but on stress analysis, weight and balance, anything and everything they threw at me. And, of course, more wind-tunnel testing. From that, I became the logical choice to be flight test engineer on the airplane when it was ready to fly.

And because I had the latest advanced mathematical training, I was given the job of analyzing the retractable landing gear for Jimmy Doolittle's Lockheed Orion 9-D, a modification of the basic Orion. That was my first contact with any of the famous early aviators who would frequent the Lockheed plant. Others included Amelia Earhart, Wiley Post, Sir Charles Kingsford-Smith, and Roscoe Turner. Doolittle, of course, was an early record-setting pilot, both military and civilian, with a master's degree and doctorate in science from M.I.T. Then he was flying for Shell Oil Company, landing in out-of-the-way fields, cow pastures, and other unprepared strips.

Retractable gear was standard for the Orion; it was the first successful application to commercial aircraft. It streamlined the plane considerably and allowed a top speed of 227 miles an hour. It was the fastest plane in service in its day and flew for Varney Speed Lanes between Los Angeles and San Francisco on a schedule of 65 minutes.

For the kind of service Doolittle required of the plane,

however, the gear needed to be strengthened. This job was tough. It required the best math I knew. To be sure the gear wouldn't come off, I doubled all the tube gauges. It cost us about 15 pounds in extra weight, but the gear worked reliably.

Every six months, Doolittle would bring the plane back to the factory to have everything tightened up. He was a hell of a good flyer and always the finest type of person. He and I are friends to this day.

When the time came for the Electra's first flight, Gross hired Edmund T. "Eddie" Allen, probably the best and most experienced test pilot of commercial aircraft at that time. The Lockheed pilots had no twin-engine experience. Allen alone flew the plane for the first time on February 23, 1934.

"Soaring gracefully into the air on its maiden flight, the sleek all-metal airliner flew easily, marking another great stride in commercial speed development of air transportation," a local newspaper reported glowingly.

The Electra was the fastest multiengine transport in the world at that time, the first able to cruise faster than 200 miles an hour. Allen averaged 221 miles per hour over a speed course shortly after first flight. Later, at 10,500 feet, the calibrated airspeed meter showed the Electra would cruise at 203 miles an hour. The plane became popular not only with airlines in the United States, but overseas as well.

In Eddie Allen, I had an excellent teacher. After the first flight, I flew with him as flight test engineer through the entire initial flight test regime—dive tests, stalls, spins, everything. It was an excellent indoctrination to the art, skill, science, adventure—all that goes into flight testing. He taught me what it was all about, what was important, what to record. And he was unflappable.

On one occasion we had to boost the Electra to its design dive speed, about 320 miles an hour, to prove that the airplane was free of flutter and control problems. We had the airplane loaded with lead bars to simulate the full gross weight. We took off from the old runway behind the factory in Burbank and climbed to about 12,000 feet. Then Eddie pointed the airplane

Establishing a foothold in the aviation industry. Kelly in his early role as a flight engineer.

down in a steep, screaming power-on dive.

At 6,000 feet when I was expecting he would start to pull out, there was a horrendous bang, and everything was flying around the cockpit. I looked over at Eddie to see what he would do, if we were going to try to jump or not. He was holding the stick with one hand, pulling back on it to bring us out of the dive, and with the other brushing insulating material from his face.

"Got something in my eye," he said matter-of-factly.

The windshield on the pilot's side of the cockpit had blown in, pulling some insulation with it and covering Allen's face. Obviously, we redesigned the windshield.

After the initial test phase, consisting of perhaps a dozen flights, Allen checked out Marshall Headle and he took over as Lockheed test pilot for the Electra. I flew with him on this and other aircraft for a good many years. When I ended my career as flight test engineer I had accumulated 2,300 hours.

I have a philosophy that those who design aircraft also should fly them—to keep a proper perspective. The engineer knows where the quarter-inch bolts may be marginal, what the flaps are likely to do or not do. I've shared the concern of the pilot. I figured I needed to have hell scared out of me once a year in order to keep a proper balance and viewpoint on designing new aircraft.

Those early experiences doubtless were important in shaping my own method of operating. A lot of engineers don't like pilots. Even more pilots don't like engineers. All of the engineers' requirements are not always met by the pilots. Engineers don't always act on the pilots' complaints. The problem essentially is one of communication.

I decided early in my career that I would try to avoid this division by doing two things primarily. I would fly as much as possible with the pilots when we had an airplane that would hold two persons, and accompany them on dangerous tests, including first flights. I've been on nine. Then, when the relatively run-of-the-mill work came along and I might not be with them to observe first hand, my door would always be open to them. They could come and tell me what they thought should be done. I didn't always follow through, but generally I found their suggestions were very, very helpful. With one exception, the pilots and I have had a fine relationship, with mutual respect, and, I think, affection.

The Electra flight test program went smoothly—until one near-disaster threatened to bring it to a halt. At the conclusion of a strenuous series of tests required for U.S. government certification, and with a representative of the Civil Aeronautics Authority on board, test pilot Headle lowered the gear for landing—but only one wheel came down. The left-hand gear would not respond.

We had just been awarded our "ticket" on the Electra, certification by the CAA, at Mines Field, now Los Angeles International Airport. Hibbard and I were driving back home in his big, red LaSalle cabriolet smoking cigars and celebrating the event.

Neither Hibbard nor I smoked ordinarily, but this occasion seemed to call for the gesture, especially since the cigars had

Lockheed's engineering department in the 1930s. Spare, close-knit, and solidly dedicated, the staff set the tone for the company's—and "Skunk Works"—efforts in the decades to come.

been presented to us in congratulation by the CAA group. We arrived in Burbank in time to witness the landing.

There was no radio communication with the plane, but the frantic waving of the group on the ground, and a check of his instruments, indicated the trouble. Jimmy Doolittle was among the observers. He volunteered to go up with a message. Gross and Hibbard dictated as Doolittle chalked on the side of his Orion's black fuselage, "Try landing at United—good luck." United now is Burbank-Glendale-Pasadena Airport. It had a longer runway than the factory landing strip, and fire-fighting equipment. Having jettisoned the lead ingots used for full-weight tests, and all extra gasoline, Headle made a beautiful landing on the right wheel only. The only damage was to a wingtip that dragged on the ground. The problem turned out to be simple to find and inexpensive to fix. The landing gear shaft had sheared; it was replaced by one twice as large.

But we didn't know that immediately. And the fledgling aircraft company couldn't afford that kind of delay in delivery—and payment. So everyone except the six of us in engineering and those in the shop required to work on that airplane was laid off, and the payroll for everyone was skipped for a few weeks—including my $83 per month. We had to get the airplane fixed and get it re-approved before it could be delivered.

First delivery of the Electra was made to Northwest Airlines in July of 1934. And paychecks were issued again.

In those early days, I was so devoted to my work and so eager to get on with it that I didn't always consider others' reactions. Hibbard had to take me out behind the barn, figuratively, for a talk several times. Once it was because I had not taken an extra flight mechanic along on an Electra test flight, and, instead, had moved the lead bars myself to shift weight in the airplane. They weighed only 55 pounds apiece, and I reloaded them with just one man, Dorsey Kammerer. He filed a complaint with the union. I had thought I was doing the right thing, saving time and money. But it had cut one man out of a job and his flight pay.

"Kelly," Hibbard explained, "you've got to learn to live in the world with all of these other people, and the sooner you learn that the better off you are going to be."

He was right. Hibbard taught me that it is much better to lead people, not to drive them. Drive yourself if you must. But not anybody else.

And Kammerer and I became very good friends. He worked for me for many years as crew chief, and a good one, on our experimental airplanes.

There was good camaraderie at the factory when I joined the force. Perhaps it was because there were so few of us, about 200 at the time of the Electra's first flight. Robert Gross played piano at the company party afterward. I had my first drink of alcohol in a celebration of that event at Neil's drugstore, a social center for us across from the factory. It was Snug Harbor bourbon. It tasted more like dregs from the bay, I thought.

For recreation, we played softball during the noon lunch period. When the engineering department became big enough, we had two teams—the engineering group and the shop team. I played left outfield and loved to hit the ball. Hibbard did some pitching. We also played some touch football.

They were a fine group, the men I went to work for at Lockheed—Chappellet, Hibbard, Robert Gross first and then his brother Courtlandt. I was fortunate to have begun my career in a company of gentlemen. Very knowledgeable they were, but also considerate and thoughtful. It was a good start. I learned a great deal from all of them over the years.

5

The Good-looking
Young Paymaster

PAYMASTER FOR LOCKHEED WAS a tall, good-looking young woman, who occupied an office right next to the other chief officers. Actually, her title was assistant treasurer, and the "front" offices at that time were on the first floor of a big, two-story, red-brick former ranch house. The engineers' room was upstairs. The first time I saw her she was working on the accounting books. Her name, I discovered, was Althea Louise Young.

She would walk through the offices and out into the factory on payday to distribute checks, so I didn't need an introduction. She would have checks, that is, when there was enough money. I remember three times when there wasn't. Everyone always was happy to see her coming!

"Althea thinks you're a snippy young kid," the chief—and only—telephone operator, Vera Doane, told me one day. I had taken Vera dancing a few times, and she and Althea were good friends. There were only four women working at the plant then—Gross's secretary, Rene Tallentyre; Alice Stevenson, secretary to Carl B. Squier—an executive with the earlier company and now vice president and sales manager; and Althea and Vera. Vera was the fourth hired because the other three all hated to handle the switchboard; she also was receptionist and secretary to Cyril Chappellet.

Well, privately, I considered that Althea may have had a

point. But I decided I couldn't let criticism like that stand, so I asked her out.

"Vera told me you were a brain," she confessed. We had a steak dinner in nearby Glendale. It cost seventy-five cents each. And we went dutch. After all, she was making twice what I was; she had been there a year before I was hired.

That evening went all right, so we repeated it. We began to go riding together, too, renting horses at least once or twice a week and riding up into the canyons in the foothills of the green Verdugo mountains above Burbank. And we both liked to dance; she was an excellent dancer.

At golf, she was much better than me. One time, I took her up on an Electra test flight. She wanted to go, and while it wasn't exactly condoned, the rules then weren't so strict about who might fly on an experimental flight. There were no passenger seats. She had to sit on the bare structure of the fuselage, but she was thrilled.

Althea was as active and alert mentally as she was physically in the sports she enjoyed. An extremely intelligent person. We found that we shared many interests and enthusiasms, had similar personal goals and ideals, and enjoyed each other's company more and more. We had fun together.

It wasn't too long before I made more money than she, and we abandoned the "dutch-treat" dates. But I wanted to be able to support a wife when I married. I didn't want her to work. Althea agreed. So it was about four years after our first date before we were married. Though we weren't regular churchgoers, we picked one we liked on Wilshire Boulevard in Los Angeles for the ceremony. We honeymooned at Yosemite in the beautiful Ahwahnee Hotel. The year was 1937.

We lived for several years very happily in a fine old house rented in the foothills on Country Club Drive in Burbank. Althea quit work and took care of the house. She was a wonderfully helpful and cooperative wife.

We continued to ride regularly every Sunday, rain or shine. After the stable went broke, we bought our horses and pastured them out west of town in rural Agoura for $3 a month per horse. Later, we rented pasturage out in Malibu Canyon and rode all around that seacoast area, back up into the hills—with the rattlesnakes—where we could look out over the beautiful valleys. Althea shared my love for the outdoors. We began to think about having a ranch of our own.

The rains in winter regularly flooded the canyons in Southern California, and after one experience on Country Club Drive—when we had to leave our house and be helped by firemen to cross a raging torrent of water, leading our dog and carrying a few personal belongings, we agreed that we would rather live on high ground.

Burbank is in the flatland of the San Fernando Valley. The hills bordering the south side of the valley offer a spectacular view of the several ranges of mountains beyond, snow-covered in the winter. Althea and I had ridden over enough country to know where we wanted to live.

In 1940, we decided that we could afford to invest in a home of our own. We bought a lot on Oak View Drive high in the hills of Encino. It was a small community then of six or eight thousand people. Ours was to be the fourth house on the hill. We discovered that Clark Gable owned the property above us and a number of other theatrical people lived in the area including Alice Faye and her husband, Phil Harris. The mayor was Al Jolson.

Our lot, literally, was in some places almost solid rock. And we wanted to build not only a house but a pool and tennis court. So, to minimize excavation in what amounted to a stone quarry, I built a scale model of the site and all construction about six feet by five feet and a foot-and-a-half tall. I put in the actual contour of the lot. Then I dug out two basements on two different levels. One would house water heater and furnace; the other, a washing machine with dryer and a wine cellar. The house itself was to be on four levels, with a 12-foot-high ceiling in the living room. Digging basements only where needed saved very expensive excavation in this rocky terrain.

Building the model was a splendid idea because I was able

to place the house exactly where I wanted it in relation to pool and tennis court.

One error crept in, though. The contractor dug the deep end of the pool an extra 18 inches deep. Well, at that point I wasn't going to change all the levels I had laid out for walks around the house and the front yard by filling in that dirt, so the pool bottom has to be the strongest in the neighborhood. It has an extra 18 inches of cement.

That pool contractor also considerably over-estimated the capabilities of his unskilled laborers. Years later we discovered, and at considerable expense, that they had not understood the use of a plumbing union, which threads two pieces of pipe together. The workmen instead used tomato juice cans to join the pool drains. They then poured a wheelbarrow of cement in each of the drain connections. It took six or seven years before the ground around the pool began to sink. Then an air hammer knocking away the cement revealed the secret of the tomato cans. The pipe had to be replaced, of course.

All that rock dug up for the two basements kept me in spare-time work for years. Though I contracted for construction of the house, pool, and tennis court, I built all the retaining walls around the property myself.

As a new homeowner in Encino, I joined the Encino Chamber of Commerce. It seemed a responsible thing for a new resident to do. Later I was elected to the Board of Directors. It was an active group and usually drew 30 or 40 people regularly to its meetings—not bad for such a small community.

Here I was to learn a lesson in how subversive groups work—not the Chamber, of course, but a group seeking to use it.

The community very badly needed a building to house Boy Scouts, Girl Scouts, Women's Club, and the activities of other worthy civic groups. I was a member of the committee to raise funds for such a building. We did, and bought several lots on Balboa Boulevard, just north of Ventura Boulevard, which runs along the foothills on the south side of the valley.

There is such a building there now, the Encino Community Building.

I proposed at the time that we of the community construct it ourselves, with enough members knowledgeable about contracting, architecture, and all the building trades. Donating time over weekends and odd hours, we could have undertaken the project without the need to raise additional money. Well, some of the people thought this was a fine idea but most did not because of the personal involvement required.

The time was ripe for what followed. At one of our evening meetings, an offer was made of whatever funds were needed if the building were to be known as the "World Peace Building."

"No," I said. "This must be a community building, for all of the various groups that need it."

Quite a brouhaha ensued, with most members of the Chamber refusing to take a position. I suspected that in a center of the motion-picture industry, where many of the most prominent actors and actresses lived, the offer of funds might have been an attempt to influence the group. Communist infiltration, perhaps?

Fortunately, I could call on Lockheed's legal counsel, a wise and experienced man, Robert Proctor of Boston, later to become legal counsel to Gen. H. H. "Hap" Arnold, head of the Army Air Corps and father of the modern Air Force.

"Bob, I think this has a peculiar odor to it," I reported. "I can't prove anything, but there is strong and very emotional talk about world peace."

"Well, you want to watch that, Kelly," he responded. "Let me check out those people and see what the connections are."

Later he called and advised, "Kelly, be sure not to call those people Communists. Don't put yourself in a position where you can be sued for everything you've got."

It was good advice. It so happened that the man who had made the financial proposal brought a court reporter to meetings to record everything I might say—that I might be sued for, presumably.

It was interesting that during this period many of my

neighbors would give me verbal support.

"We're with you."

"You're doing fine."

"Keep it up."

But they also told me: "I have a store down here."

"I have a lumber yard."

"I don't dare to antagonize people here; it's bad for business."

I was mad as hell, because they all agreed that there was something fishy about that offer of money. It wasn't accepted, finally. Well, it all ended with me not being re-elected to the board of the Encino Chamber of Commerce.

About 20 miles west of our home in Encino was undeveloped ranch land, where Althea and I pastured our horses. Several years after building our home, we had the opportunity to buy the Lindero Ranch, 226 acres of rolling country with a stream bordering its west side. Lindero means line or boundary and was the northwest corner of an original Spanish land grant.

We built a small house on a mountain top with a view of the Pacific Ocean to the southwest and a range of mountains in the other direction.

My early experience in construction was invaluable. With no utilities in the area at the time, we had to provide our own water, light, and power.

The house we built ourselves was of concrete block, about 900 square feet. A large living room with six-foot high windows across about 30 feet of the front provided a magnificent view. A huge fireplace heated the entire structure.

For electricity, we installed a small power plant that ran on cheap propane and also powered our stove and refrigerator. We had all the comforts.

Other jobs were to dig a septic tank and put in a water tank. That was challenging because I had to move by tractor a 1,000-pound water tank slung on a boom downhill to a position above the house. Before I began this operation, I carefully computed tilt angle of tractor and tank, to keep the whole setup from tumbling downhill and through the new house. With a wind-mill at the bottom of the hill pumping up to the tank, we had a fine water system.

Althea and I ran the ranch by ourselves. She did as much work as I—such as driving tractors and taking care of the horses. She became very skilled at operating the tractor and often would mow or rake hay while I was at work. It was a real partnership.

We planted 110 acres of oat hay, plowing and discing the fields ourselves. We harvested the baled hay, too.

Althea loved animals and convinced me that we should buy a herd of registered cattle, so we invested in about 20 Herefords. She enjoyed keeping the breeding records and everything else about ranch life. We both loved Lindero. As a pet we had a scrawny calf we called "Hardly Nothing" and re-named "Almost Something" as it grew.

One day I forgot that Althea and I were equal partners! We were discing in preparation for planting oat hay. I didn't think she was handling the C-2 caterpillar tractor properly, dragging the brakes and making hard use of the clutch. I hate to see any kind of machinery not treated very gently. So I corrected her in what I considered to be my most constructive manner.

"If you don't shut up," she responded, "I'm going to run over you with this thing!" And she headed for me at full speed—about four miles an hour. I quickly departed the scene to let things cool down and she finished her discing.

One great joy was riding horseback over our own land and surveying our crops. While we bought more horses for the ranch, we still had our first two, Prince and Mac. Both lived long lives, each more than 27 years, although we quit riding them long before that age, of course.

Althea and I never had children, but it was not from lack of desire. We were very compatible and had a good and happy life together.

6

Wiley, Amelia, and Others

Newspaper headlines in August, 1935, saddened a nation that had loved a famed humorist-philosopher and admired a brave, pioneering aviator.

"WILL ROGERS AND POST DIE IN AIRPLANE CRASH," bannered the *New York World-Telegram.* "Pair Forced Down by Faulty Motor; Fall in Takeoff."

"ROGERS-POST ARCTIC DEATH CRASH RELATED" read the headline across the front page of the *Los Angeles Times.* "Plane Hits River as Engine Falters While Taking Off." A separate story related, "Fog Blamed for Death."

Wiley Post was one of the first of the early famed flyers I met when I began to work at Lockheed in 1933. He already had flown around the world once, with Harold Gatty, in 1931. He became the first man to make the flight twice in a challenging solo effort in July of 1933. Both flights were made in a Lockheed Vega, the famed *Winnie Mae.* Wiley earlier had been a Lockheed test pilot. And, before that, a roustabout in the oil fields of Oklahoma. That's where he lost an eye, resulting in that identifying black eye patch.

It always was a mystery to the rest of us how anyone could fly so well and seem to have depth perception with just one eye. But he could, and he demonstrated his courage and excellence as a pilot many times.

When I met him, Wiley wanted to attempt high-altitude flying, convinced that there was a future in the stratosphere where higher speed would be possible because of less air resistance and good tail winds.

He had persuaded the Bendix company to furnish a very high-pressure supercharger for his Vega's engine. He planned that it would furnish not only engine air but also air for a pilot's pressure suit he had developed—it looked like a deep-sea diver's outfit. This was to make possible flights at high altitude where there was little oxygen. He wanted to reach 40,000 feet and fly across the country at 400 miles per hour, from west to east.

To do this, he wanted a special skid installed under the belly so that he could drop the landing gear after takeoff to save weight and improve the aerodynamic performance.

With Jimmy Gerschler, then assistant chief engineer, I worked on the droppable gear, on the skid, on the air intake, and on bettering performance. But the best I could calculate for cruising speed in level flight was nearer 260 miles an hour.

After work, Wiley and I would assemble at Neil's drugstore across from the Lockheed factory, a popular meeting place. Wiley would have a beer and I'd have a sherry. He kept after me to prove that he could exceed 400 miles an hour average cruising speed. I managed to get close to 300 mph, but never better than that.

"Kelly," he promised, "if I do get across and average 400 mph, you're going to have to buy me a case of whiskey. And if you're right, I'm going to buy you a 20-inch slide rule to replace that little 12-inch one you use, so you'll get better numbers."

I never got the slide rule. And he didn't get across country.

He tried three times. You'd think after once or twice he'd have given up, but not Wiley. He would take off from the Burbank airport and drop his gear at the end of the runway. But after being airborne he'd have trouble with the engine and have to make a forced, unprepared landing on the skid. He'd reach the dry lakes of the Mojave Desert, which made good emergency landing fields. This required all his pilot's skill and depth perception. Once, he got as far as Cleveland. He had averaged 253 miles an hour at an altitude of 30,000 feet, but was forced down again. He couldn't use all the power he had without over-boosting the engine.

But it was a very good plan; and if the engine had held up Wiley would have been all alone, way ahead of everyone else. No one could have broken that record. He had a cleaner airplane than the Orion, last of the old Lockheed plywood aircraft and the fastest commercial ship of that day. It was capable of 226 miles an hour top speed and was faster than some military models. Wiley had a thinner wing, lighter weight without the landing gear, more powerful engine, and twice the cruising altitude—in thinner air offering less resistance.

Wiley pursued yet another project. He had modified an Orion, substituted wings from another Lockheed plane, the Sirius, and had floats attached for water landings. He and his friend, Will Rogers, planned to tour the Yukon, Northwest Territories, and Alaska to do some hunting and take a leisurely vacation. Post had a passport for Siberia and there was talk of taking the plane around the world.

But Wiley also had put on this airplane the biggest engine he could get, with the biggest propeller and a different gearbox.

"Wiley, you'd better watch this," I cautioned. Gerschler warned him too, "You're getting out of balance; the airplane is too nose-heavy with this big prop."

"Oh, I'll handle it, I'll handle it," Wiley countered.

I persisted. So did Gerschler.

"You'll have trouble on takeoff," I argued, "because I doubt that there is enough elevator power to get the nose up."

But he got the plane certificated by the CAA and flew off.

He managed to get the nose up for takeoff by rocking the airplane fore and aft on its floats with power on until finally it would bounce up into the air, and he was airborne.

In the very poor visibility in which he and Will Rogers attempted that last takeoff, it is doubtful he actually had a horizon visible. Under those conditions, a pilot can lose his sense of reference, cannot tell his angle of attack. And then to pull up at too high an angle and stall or have the engine fail . . .

Wiley was an exceptional pilot who had overcome many obstacles. He and Rogers had covered a lot of territory to the northernmost point of Alaska, 300 miles within the Arctic Circle, before the fatal crash. According to newspaper reports, they had stopped for three hours at an Eskimo encampment to repair a faltering engine. The engine failed again on takeoff and they crashed on the frozen tundra. Wiley's wristwatch stopped on impact: 8:18 p.m. An Eskimo ran the 15 miles to Point Barrow to report the tragedy. Famed bush pilot Joe Crosson flew the bodies to Fairbanks, where Col. Charles Lindbergh, as a director of Pan American Airways, personally directed their return home.

Tragic headlines of another day—these in 1937—recounted the story of Amelia Earhart's disappearance on an around-the-world flight.

"MISS EARHART MISSING OVER PACIFIC: ONLY A HALF-HOUR'S FUEL, NO LAND IN SIGHT, SHE RADIOS—THEN SILENCE." Four columns with photo, maps, Earhart's last dispatch to the paper, and the news story covered the front page of the *New York Herald Tribune*.

"AMELIA EARHART LOST IN PACIFIC; RADIO FLASH-ES FAINT SOS," reported the *Los Angeles Times*. "Plane Joins Ship Hunt for Flyers." Ironically, in light of later speculation about Amelia's disappearance, that same front page of July 3 reported, "Russian-Japanese Crisis Eased by Troop Withdrawal."

America's most famed woman pilot of that time, Amelia Earhart, and her navigator, Capt. Fred Noonan, were not heard from by Pacific area listening posts after their radio went silent at 11:12 a.m. Pacific Standard Time on July 2. Faint signals were picked up at 1 a.m. the next morning by a ham operator in Los Angeles and by a steamship several hundred miles too far away to give assistance.

The flight around the world, begun "just for fun," in Earhart's words, ended in silence that day.

Amelia already had established a distinguished career by the time I met her. She was the first woman to fly the Atlantic—

as a passenger in 1928, solo in 1932—and the Pacific—Honolulu to Oakland, Calif., in 1935. She set three other impressive records in her Vega during 1935: Burbank to Mexico City in 13 hours, 32 minutes; first non-stop flight from Mexico City to Newark, N.J., in 14 hours, 19 minutes; and a transcontinental record for women, Burbank to Newark, 13 hours, 34 minutes, 5 seconds.

In her book, *The Fun of It* published in 1932, she wrote of her solo transatlantic flight: "It was clear in my mind that I was undertaking the flight merely for the fun of it. I chose to fly the Atlantic because I wanted to. It was, in a measure, a self-justification—a proving to me, and to anyone else interested, that a woman with adequate experience could do it." And in helping to "put the theories to practical use . . . (toward) efficient flight," this early feminist wrote: "That women will share in these endeavors even more than they have in the past, is my wish—and prophecy."

When I met her, she had a Lockheed Model 10E; the original Electra with slightly more powerful engines—550 horsepower Wasps instead of the original 420 horsepower Wasps. Her ambition was to fly around the world.

The work I did with her basically was to find out how to get the absolute maximum mileage out of the plane for the around-the-world flight. The two of us, she as pilot and I as flight engineer, would fly her Electra with different weights, different balance conditions, different engine power settings, different altitudes.

We had in those days a gadget known as a Cambridge analyzer to analyze exhaust gas; you used it repeatedly, resetting mixture control and leaning out engine fuel to get maximum miles per gallon. Amelia learned how to do this, too. She also had the advice of another aviation veteran, the famous racing pilot Paul Mantz, on installation of fuel tanks, instruments, and other special provisions for the flight.

Her original intention was to fly around the world to the west, and I listed fuel loads recommended for the first six long hops in a letter to her dated February 17, 1937.

"Dear Miss Earhart," my letter began. "The following fuel loads are recommended for the flights noted. These figures are subject to change depending on actual fuel consumption tests which you are going to make, as I discussed with Paul Mantz. A 25% margin (assuming zero wind) for range is included in the following figures."

Then I listed distance, fuel load, and gross weight for each leg of the trip, from San Francisco to Natal.

"The use of 10° to 30° of wingflap with takeoff power will reduce the takeoff run about 20%. If a normal, good runway is available, with a length of 3000 feet (for the heavier loads), no wingflap is required or recommended as the ship will take off in 2000 feet with a load of 14,000#. The greatest danger in using wingflaps on takeoff lies in the reduction of directional control at the beginning of the run, and in retracting the wingflaps after takeoff. The flaps *must not be raised* unless an airspeed of 120 m.p.h. is reached, in order to prevent losing altitude as the flaps retract.

"If the runway is rough, so that the landing gear must take a terrific beating during takeoff, some flap (15°) should be used.

"During all takeoffs, the airplane should be held with the *tail up* in approximately level position in order to get the best possible takeoff. Lift the airplane from the ground as soon as it is safe to do so, to relieve the load on the landing gear. After leaving the ground, hold the ship *low* until a good margin of speed is obtained before starting the climb. This procedure makes use of the ground effect on lift and drag in the best manner.

"When a takeoff is made with flaps down, the procedure after leaving the ground should be to retract the landing gear first (soon as possible) and then climb to a safe height with flaps down, level off and accelerate to a speed of 120 m.p.h. Retract the flaps, maintaining an air speed of 100 m.p.h. or more. Resume climb at normal power when flaps are up.

"The above procedure is fairly complicated; so it is generally recommended that no flap be used on takeoff unless necessary.

With her famed Electra—Lockheed's first all-metal aircraft—as backdrop, aviatrix Amelia Earhart discusses coming flight with Kelly Johnson.

"Yours very truly, Clarence L. Johnson."

On her first attempt at the around-the-world tour, she ground-looped the plane in Hawaii and stripped off the landing gear. The plane had to be returned to the factory by boat. We had more discussions, of course, about how to prevent ground looping; and it did not happen again.

But she changed her direction for the second attempt, flying instead toward the east. I don't know why. Perhaps because she expected more favorable winds. She almost made it. Leaving Oakland on May 30, she flew across country to Miami, then to Puerto Rico, down the coast to Brazil, across the Atlantic to Africa, to India, Australia, and Lae, New Guinea. The next to the last leg, the longest was Lae-Howland. Then it would be Honolulu and on home to Oakland.

With Howland Island, Earhart and Noonan were trying to hit a very, very small speck in the broad Pacific Ocean, an island one and a half to two miles long and rising only about two feet above water. Fred Noonan was a very good navigator, but it became apparent from radio conversations recorded by the U.S. Navy and others that the sky was so overcast that they could not get down to see any checkpoints nor high enough—above 20,000 feet—to get "sun shots" to check their navigation.

They had been in the air for 23 hours and, so help me, that's all the time they had fuel for. They did not know their location when they sent their SOS. The Navy mounted a tremendous search effort and attempted to locate them with direction finders but couldn't. The two had a rubber dinghy with them, and, if undamaged, the plane could have floated with its gas tanks emptied. I am convinced that they attempted to ditch the airplane and didn't get away with it.

There has been speculation since that Amelia was on a spy mission to overfly the Japanese and photograph buildups of military facilities and operations. I doubt this. The only camera she had to my knowledge was a Brownie. And there were no openings in the aircraft that would have permitted good aerial photographs.

Also, it has been implied that Amelia may have been a poor pilot. She was a good one when I knew her. She was very sensible, very studious, and paid attention to what she was told. In person, Amelia was kind, gentle, quiet in speech and manner. She *was* the "Lady Lindy" she was called.

I have always had a great admiration for Amelia Earhart as a lady and as a pilot.

Sir Charles Kingsford-Smith, the Australian, was another pilot I got to know well. I worked closely with him, as I did with Amelia, on the optimum way to get the most miles per gallon for his Lockheed Altair. He already had set distance records in Australia and internationally—Oakland to Brisbane in 1934. In 1935 he wanted to fly from England to Australia. We were preparing for that project.

His little Altair didn't have an opening in the bottom of the fuselage through which I could drop instrumentation, so I would stand up in the rear cockpit, letting an airspeed "bomb" trail below. This was a heavy lead weight with a pitot tube on it—to measure airspeed—which we let down some 100 feet below us, out of the wake of the airplane. It gave a very accurate reading. We made a number of such flights from Burbank out over the Pacific, calibrating airspeed and determining best operating details for maximum range.

On November 6 of that year he and his copilot-navigator, Thomas Pethybridge, left London for Sidney. On November 8, they disappeared en route to Singapore. We theorized that they had struck a cliff or mountain while at low altitude, damaged the plane, but were able to continue flying a short distance.

A wheel and tire were found in the ocean, but the flyers and the aircraft never were. It was a loss to aviation and a personal loss to me. Many years later, I was to receive an award from the Royal Australian Aeronautical Society for my work with Kingsford-Smith.

Col. Roscoe Turner, dapper always in puttees and boots or impeccably tailored suit and bowler, may have been the most colorful of that early group. He had been a lieutenant in the Air Service during World War I, was an officer in the Civil Air Patrol, holder of the DFC, both the Thompson and Harmon Trophies, a winner of Bendix and National Air Races, and a barnstormer and stunt pilot. He flew his Air Express to the factory accompanied by his mascot, Gilmore, a lion.

He had acquired Gilmore as a cub and named him for his sponsor, Gilmore Oil Company. Now Gilmore was a very nice lion, and used to roam the factory at will, but he got to be darned big and potentially dangerous. He must have weighed several hundred pounds, and Turner kept him on a chain during visits to the factory. But once he got loose and chased Althea up the stairs of the main office building before Turner could recapture him. The lion was just playful, but at 400 pounds!

Other early pilots? They were in and out of the plant frequently—Charles and Anne Lindbergh, Laura Ingalls, and Ruth Nichols. There was no security problem at the plant then. Aircraft owners and visitors just walked right in.

The only woman pilot with whom I worked aside from Amelia was Laura Ingalls. The two women could not have been more different. Laura had an Orion fitted with wingtip fuel tanks and other modifications for flying cross-country and setting speed and distance records. She did, in both directions, in 1935. She also had a Lockheed Air Express in which she had made a 17,000-mile solo tour of South America a year earlier—the longest trip by a woman and the first for a woman around that continent and over the Andes. She was a neighbor of ours—Althea's and mine—on Country Club Drive in Burbank, and I worked with her on her Orion.

Laura spent a lot of time at the factory inspecting her airplane and was a stickler for detail. But she had one annoying habit. She would quiz the mechanic about the airplane's performance, and ask me how to change a spark plug. She cross-checked on everything with everyone.

Col. Charles A. Lindbergh flew an early polar route survey for Pan American Airways in 1933 in an early Lockheed Sirius modified with pontoons. I did not work with him on this plane, a project handled by Jimmy Gerschler. I did not meet the Lindberghs until many years later.

Lindy encountered terrible flying conditions—fog, ice, and generally bad weather. Landings were dangerous when the lakes were iced over and he had to set down on the ocean. The survey didn't result in an immediate commercial polar route, but it did inspire Anne's book, *North to the Orient*.

On the occasion of a banquet in Beverly Hills many years later, attended by the Lindberghs and Althea and me, I discovered that my dinner partner would be Anne. She is a serious writer in her own right, a sensitive poet, and a concerned advocate for aviation. After we had covered our mutual flying interests, including incidents from their long-range flight in the Lockheed seaplane, we got around to discussion of what issues we thought were really important in life. When she asked what they were for me, and I told her, she agreed. They are, if you are interested:

1. Belief in God. When you're in a difficult situation, ill, or in danger, and wonder if you're going to make it through to the

next morning, your faith is one of the most important thoughts that will sustain you.

2. Health. Without it, one cannot be truly happy. It certainly has been important in my life.

3. Purpose. We must have a purpose in life, doing something we want to do and doing it well. This will provide the necessities—security, money, and rewards of other sorts.

4. A wife or husband who loves and understands you.

5. Respect of the people for whom you work and who work for you.

That pretty well sums up my philosophy of living, and I was glad to have the concurrence of such a person as Anne Lindbergh.

7

A Family of Aircraft

THE ORIGINAL LOCKHEED ELECTRA begat a line of characteristically twin-tailed commercial transports. The first of these was the Model 12, followed by the Model 14 and an advanced experimental model—one of a kind—the XC-35, which won for the U.S. Army Air Corps the Collier Trophy for pioneering high-altitude flight. The Model 14 led to Lockheed's first airplane for World War II.

The Electra became a very successful airplane—with the airlines, military, and private customers.

It was comfortable, carrying five persons on each side of the aisle, and had a roomy cockpit for pilot and copilot. It was the first airplane with an all-metal surface to go into production in the United States. And it was fast—190 miles an hour cruising speed.

The first derivative was a smaller version, Electra Jr., or the Model 12, for customers who didn't need or couldn't afford the Model 10's capacity and performance. It carried only six passengers, was somewhat faster at 206 miles-an-hour cruising speed, and, at $40,000, cost $10,000 less. Introduced in 1934, it set a number of new world speed marks for transports. Production didn't halt until 1942, and then only because facilities were needed for military projects.

Next came the Model 14 Super Electra—and it was. It had the latest advances in everything to do with aircraft design and manufacture. When it first flew in 1937, it was the fastest, at 237 miles an hour cruising speed, 257 top speed, of any commercial transport in the United States. It beat the DC-3's flying time from Los Angeles to New York City by four hours.

It used the newest Wright Cyclone engines, the newest and strongest aluminum alloy then available, a new Plexiglass for windows. And it had two totally new innovations: the Lockheed-Fowler flap which significantly increased wing area for takeoff and landing, and to direct airflow, "letterbox slots," so-called because that's what they resembled.

The third Electra derivative was developed in secrecy for the U.S. Army Air Corps. The XC-35 was the world's first successful pressurized substratospheric plane.

Wiley Post's work in 1935, with his pressure suit and supercharged engine, had proved that man and machine could operate above 30,000 feet.

That next year, the Army Air Corps gave Lockheed a contract to modify an Electra cabin to hold air pressure—a 10-pound-per-square-inch differential between inside and outside. The extra weight required to strengthen the cabin was 1,486 pounds. The XC-35 flew and was delivered to the Air Corps in 1937, and won the Collier Trophy for having made the most valuable contribution to aircraft development during that year.

My own experiences with high-altitude flight began with Wiley Post, but another involvement during Electra flight testing with our chief test pilot, Marshall Headle, led me to direct my attention very early to this field. We needed to demonstrate to the Brazilian government that the airplane it had purchased could climb to an altitude of 23,000 feet. And they wanted proof, not just the word of a flight test engineer or pilot. They required that we fly to that altitude with a sealed barograph, an instrument that automatically records barometric pressure.

At that time, the prevailing opinion was that you should not breathe any more oxygen than necessary at altitude because of the threat of oxygen poisoning. And, of course, all we had to breathe with was a cigarette holder connected to an oxygen line. It took us three flights before we reached the altitude requirement. We'd climb and climb and not make it. Finally we changed airspeeds and some other factors and managed to reach the desired altitude.

But when we landed, I felt so ill that someone had to drive me home. I literally fell onto the bed and practically had to hold on to keep from falling off. I was really, really sick. It was a frightening experience, and sparked my continuing interest in oxygen systems, pressure suits, and pressure cabins from that day to this. I've had ample opportunity since, in the light of later developments, to respect those early pioneering tests of Wiley and others.

Some of our other test equipment in those days was as primitive as that cigarette-holder oxygen system. But it worked. To determine the drag of the tail wheel in flight tests on the Model 12A, I rigged a standard fish-market scale to the wheel and strut—just in front. The drag force would register on the scale by causing the arm to move. In this case, it showed that the drag was not important enough that we need make the gear retractable.

The first flight of the Model 14 was one I won't forget. Marshall Headle was pilot and I was flight engineer, having worked on design of the new wing flaps and a good deal of the rest of the airplane. It was an important flight. No one had been able to put the Fowler flaps on a commercial airplane success-fully. They weren't the usual wing flap that was lowered to act as an air brake. They slid backward out of the wing and added effectively to wing area, allowing a large wing for landing and takeoff control and a small wing for speed in flight.

Lockheed didn't yet have its own wind tunnel, but I had run a great many tests at California Institute of Technology (Cal Tech) in Pasadena, where Lockheed and six other aeronautical companies would rent tunnel time. I had set the design condi-tions for installation of the flaps in the airplane and was confi-dent that I knew something about it.

We took off from the old runway behind the factory, and when we got to altitude and started to lower the flaps for the first time to test their operation, there was a "bang." The flaps went all the way down and we couldn't raise them. We had lost hydraulic pressure in the system—not a serious defect but critical at the moment. We experimented with different approaches to getting the airplane back on the ground and discovered that the slower we flew the more the flaps came up. Not too healthy a situation for a landing—at least, not with the limited runway length we had.

But I happened to recall, fortunately, that at about a 20 percent flap setting there was a bend in the flap tracks that would stop the flaps. So we went on in for a landing with flaps free; and as we slowed and they came to that 20 percent setting, they held that position and we landed safely. Most of our Lockheed aircraft now have triple or quadruple redundancy in hydraulic as well as other key systems.

At the time of the Model 14 flight test program, I still hadn't learned my lesson about not taking on more than I was authorized to do. We had a lot of work to complete and I wanted to get on with it. On one Sunday when our chief pilot Headle wasn't available, I prevailed on another pilot, "Mac" McCloud, to fly the plane while I went along as flight engineer. He was a licensed pilot but hadn't been checked out in that airplane by a qualified pilot.

I had no license to fly the airplane, but I knew very well how to fly it since I'd been on every flight from the first. I gave him takeoff speeds, direction on handling the flaps, and power settings for the engines. If all had gone well, no one would have known about our unofficial status.

The flight itself went off just fine. I gave directions for landing—hold up the tail, put the nose gear down first, then let the tail settle. The landing also was fine until we got about 800 to 1,000 feet down the runway. The airplane suddenly yawed to the right and ended up going sideways. I looked out my side of the cockpit and there, sticking up through the right wing was the main landing gear strut! Migawd, I thought, here I've taken an airplane, checked out a pilot illegally, and wrecked the plane. There goes my job.

But when the inspectors checked they found that, instead of six bolts holding the landing gear, only three had been installed when it was signed off by inspection. McCloud and I were in the clear.

A very, very serious danger in those days was icing, because few airplanes could provide enough heated air for the carburetor to prevent ice formation in it. And with ice in the carburetor, the engines would lose so much power that you were in real trouble even if they didn't die completely.

For the first time, with the Model 14 we had available a carburetor designed to correct that—the Chandler-Evans non-icing carburetor. We decided to incorporate it in the airplane. Fortunately, we didn't rely on it solely.

One of the tests that had to be passed before the aircraft could be certificated with this modification by the CAA—now the Federal Aviation Administration—was performance in icing conditions. The local CAA inspector, Lester Holoubek, was aboard to see how the new carburetor operated on one flight out of Mines Field. We had to fly up through about 3,000 feet of heavy undercast and had just broken clear for another 1,000 feet when the left engine gave a few hiccups and quit. It wasn't long before the right engine slowed and gave every evidence that it would follow suit.

It was a scary situation, especially since we couldn't see the airport and knew that we had 3,000 feet of icy clouds to drop through on the way down—with the CAA inspector aboard. But we had not relied totally on the new carburetor and had provided for alcohol injection, too. We turned on a small, hand-operated pump and quickly dissolved the ice before we had descended below 1,500 feet. The engines operated again and we were able to maneuver down out of the overcast and land. We found we not only had a problem with the non-icing carburetor but with the CAA inspector, who decided—quite rightly—that if we were going to fly with that carburetor, we had to prove it with a lot more flying in icing conditions.

The problem was compounded when about that time one of Northwest Airlines's Model 14s, flying between Seattle and Minneapolis-St. Paul in icing conditions, crashed near Bozeman, Mont. The cause of the crash, of course, wasn't immediately known. I was on the investigating team and joined in inspecting the wreckage.

It was apparent immediately that the two vertical tails were not on the airplane—they were just gone. Absence of the tails, of course, would lead the airplane into a very unstable flight regime and inevitable crash. Heavy snow was on the ground and we accepted the offer of the Bozeman ski club to hunt for the missing tail pieces. I even joined the search on skis one time—profiting by my boyhood experience in the winters of northern Michigan.

When the tail was found and brought to the accident investigation center, the rudder was missing completely. Just the control to the rudder tab was hanging there, with nothing behind the hingeline.

The regulatory agencies immediately imposed requirements for lead balances on both vertical tails above and below the horizontal stabilizer as well as some other specific changes.

But I wondered if we weren't overlooking something. I had seen the tab control, the ball bearing that is supposed to keep the tab in proper position as the rudder turns left or right. The bearing had been broken and there were no balls in it. The whole center race—the track the ball bearings ride in—was gone. And so was the tab—the movable trailing edge of the tail rudder.

Back at the plant, I convinced Hibbard and others that we should run a tunnel test on a full-scale vertical tail and find out what conditions would cause flutter. We built our own wind tunnel in 1939, the first sophisticated one in private industry. But at that time we were able to use the Guggenheim wind tunnel at Cal Tech under the distinguished Dr. Theodore von Kármán and Dr. Clark B. Millikan. The test section was a cylinder ten feet in diameter, obviously a very difficult space in which to work when changing models, but the tunnel had enough capability to exceed by far the speed at which the Northwest aircraft was cruising when it was lost.

That tail wouldn't flutter in the wind tunnel no matter what we did. But then we disconnected the tab, simulating the broken bearing, and immediately the rudder blew off the tail assembly. We went ahead with the lead balances decreed by the CAA; but the real cause of the accident in my opinion was that someone, a mechanic on the production line or in the airline's overhaul shops had not held the bearing properly when he adjusted the tab setting and had cracked the race but not broken it completely. The break occurred later in rough air and caused immediate violent flutter. I had never seen such violent flutter as with the simulated condition in our wind tunnel.

We still had to conduct 50 hours of flying in icing and rough weather to satisfy CAA conditions. And it had to be in the same air corridor where the plane had been lost. So Headle, Holoubek, and I headed for Minnesota. We stayed on the ground when the weather was nice and everyone else was flying, and took off in the worst of it.

We would fly into the roughest air we could find to prove stability of the aircraft, and in the worst icing conditions to prove the new non-icing carburetor. We'd have two to four inches of ice on other parts of the airplane, but the engines kept running. On one of our most exciting flights, we collected so much ice in just four minutes that with both engines on full power our indicated airspeed was just 90 miles an hour. And we landed with full power on.

I was so impressed with the rapidity with which ice can build up and its severe effect on aerodynamics and control that I was prompted to write one of my first technical papers on the subject for the benefit of others. ("Wing Loading, Icing and Associated Aspects of Modern Transport Design," Journal of the Aeronautical Sciences, December 1940). To this day, the only thing I fear more than ice is hail. A separate icing wind tunnel later was built as another research facility at Lockheed. Pilots who haven't experienced it do not realize that it takes damned little ice to cause a horrible crash.

Another little detail we had to prove was that the control cables would not go so slack because of the low temperatures that they would allow flutter under certain conditions. To do this we had to measure elevator cable tension, and the only place to reach the cable was by removing the toilet and reaching down through that space to attach a tensiometer. This was Holoubek's job as inspector. One day while he was checking this instrumentation, we hit a particularly severe bump. I still can see his feet sticking up through the opening as he yelled for help to get out of there.

The Model 14 had so much power with those new Wright-Cyclone engines that this actually produced a problem. With so much of the inboard wing in the propeller slip stream, it was almost impossible with power on in flight to stall the middle of the wing. So the wing, if it stalled, would stall—that is, lose lift—on the outboard end near the aileron, the trailing edge flap used for lateral control. This was not good, especially if one wing tip stalled before the other. It would cause the plane to roll badly.

All sorts of corrections were considered, including change of the wing shape itself, to control local stall characteristics. In the wind tunnel, we put hundreds of tufts of yarn on the wing so that we could watch stall patterns develop in simulated flight. When would air flow separate and under what conditions? It wasn't the first time we had used yarn to observe air flow, but it certainly was the most complete such test we ever had conducted.

And for the first time we had an AO, automatic observer. A complete set of instruments was connected to the airplane's systems. One camera would show the time and all flight conditions—airspeed, altitude, rate of roll—28 different recordings. This was synchronized with two other cameras that took pictures of the left and right wings when a stall developed.

That test program probably was the most thorough for specific aircraft performance characteristics undertaken to that time. And then in flight I joined the pilots in making 550 stalls and falling all over the sky in the San Fernando Valley for several months. It was an interesting way to make a living.

But it was not possible then to make these tests in the wind tunnel because we could not simulate the slip-stream effect of the propellers. When the electric motor models were scaled down to fit into the proper-scale engine nacelle they simply could not produce enough power for a realistic test. We had no choice but to fly.

The result of this work was the "letter-box" modification already mentioned. A venturi-like opening at the bottom of the wing narrowed as it carried air through the wing, so that it was released over the top surface at much higher speed than normal at that point on the wing. And it was fresh air, not tired of flowing over half the wing. It was able to do this at angles of attack much more successfully than could any wing section alone. There were at the time other retractable wing slots— notably by De Haviland—but we found the complexity and maintenance problems not tolerable.

What we did was based on the Coanda effect, named for the French engineer who discovered that if he blew air on a curved surface, the flow did not separate from it but tended to remain stable on that surface. The principle is used routinely now in boundary-layer control.

These were good times for me personally. In 1938, I became chief research engineer for Lockheed.

When Lockheed's engineering department began to expand I recruited some of the students I knew from the University of Michigan. Willis Hawkins was first; I had corrected his papers for Professor Stalker and knew his scholastic ability. Rudy Thoren and John Margwarth followed. Carl Haddon, who had been a year ahead of me in college, joined us.

It was almost like a university club. And then in night classes at Cal Tech I met another group of young engineers. Phil Colman was recruited there. With Irv Culver and E. O. Richter, these were the stalwarts I started with as chief research engineer.

The work on the "Lockheed-Fowler Flap" brought me my first major award as an engineer, the Lawrence Sperry Award for "important improvements of aeronautical design of high speed commercial aircraft", in 1937.

The two major developments arising out of the Model 14 design, the Lockheed-Fowler flaps and the "letter-box" slots—both of which give the airplane excellent handling characteristics—were to become especially important in light of the unexpected role this airplane was to play in history.

8

War and Mass Production

THE YEAR 1938 WAS TO CHANGE LIFE FOR US at the growing
Lockheed plant as, indeed, it did for everyone around the
world. Hitler was on the move in Europe, and the British
military—despite their prime minister's assurances of "peace in
our time"—could see the inevitability of war. That flawed mission of Neville Chamberlain's meeting with Hitler in Munich
was flown, incidentally, in an Electra. That airplane and its two
commercial derivatives had met wide acceptance.

Remembering the lessons of World War I, the British knew
that in event of war with Germany they could expect tremendous shipping losses. They needed, among other things, an
antisubmarine patrol plane. In April of that year, they sent a
purchasing commission to the United States to buy a training
plane and a coastal patrol bomber.

The committee was not even scheduled to visit the Lockheed plant. But the company officers were informed via tele-
gram by the British air attaché in Washington just five days
beforehand that the group would come to California. And they
decided to do something about it. Our Model 14 was fast, about
the right size, and capable of carrying the necessary equipment. We hurriedly built a full-scale wooden mockup of an
antisubmarine version.

Kenneth Smith, then working as a sales representative for
Robert Gross, studied newspaper photos of the group and
memorized their names. When they landed at Glendale airport, he greeted each and invited the group to inspect the

mockup Lockheed had developed. They did—that very day.

Our mockup, of course, was only our guess as to what the British would need. But when they saw how enthusiastic we were about the project, they gave us a better idea of what they had in mind. Their visit was on a Friday. We incorporated what changes we could over the weekend and called them on Monday for another inspection. And we had prepared reports to show the plane's performance. They were so impressed that this little company had the gumption to address their problem that they invited us to England to talk to their technical people.

Robert Gross's younger brother, Courtlandt, who had by now joined Lockheed management, led the team. The other three were our lawyer from Boston, Bob Proctor; Carl Squier, then vice president for sales; and I. We sailed on the *Queen Mary.* Courtlandt has insisted that I found this means of locomotion very inefficient and mentally redesigned the ship on the way over. It sounds typical, and I suppose I did.

At the Air Ministry, our proposal lasted about 30 minutes. We had made the mistake of basing our mockup on stacking the bombs and torpedoes in the manner of the U.S. Army Air Corps in racks from floor to top of cabin. The British wanted everything in the bomb bay. They wanted to install their own oxygen system and other equipment so that everything would be supportable directly from their stores. And they wanted a gun turret to protect the plane from the rear and also forward-firing guns. There wasn't a powered turret that would fire in any direction in the United States at that time. All of these changes affected the entire structure of the airplane—weight, balance, performance. This required almost a complete re-design and we decided to undertake it on the spot.

Following the meeting, we bought a drawing board, some T-squares, triangles, and other drafting equipment, and headed back to our quarters in Mayfair Court. I had to fit in all this new equipment, re-arrange copilot and radio operator positions, make weight and structural analysis, figure contract pricing, and guarantee that the design would meet certain performance requirements.

It was a three-day holiday weekend—Whit Sunday, Whit Monday. I worked a solid 72 hours on this redesign—not taking time for sleep, just catnapping briefly when absolutely necessary. I was a rumpled figure.

When finally I fell into bed for some very sound sleep—in the room I shared with Courtlandt to save on expenses—it was the first time I had removed my clothes in 72 hours. I awoke the next morning to discover that he had had my suit pressed and my shoes shined. How wonderful, I thought, that the head of the company would do something like that for an employee. That kind of consideration was typical of the gentlemen I worked for.

When we reappeared at the Air Ministry with a complete new layout on Tuesday, the British were surprised that we had worked through the holiday. In another week or so of meetings with the British, we were able to answer most of their questions. But they had one for Court Gross. He was called aside by the Chief of the Air Staff, Air Marshal Sir Arthur Virnay, and asked, as best Gross remembers:

"Mr. Gross, we like your proposal very much, and we very much like to deal with Lockheed. On the other hand, you must understand that we're very unused in this country to dealing—particularly on transactions of such magnitude—on the technical say-so of a man as young as Mr. Johnson. And, therefore, I'll have to have your assurance, and guarantee, in fact, that if we do go forward, the aircraft resulting from the purchase will in every way live up to Mr. Johnson's specifications."

Courtlandt assured the Air Marshal that he and his brother, Robert, had "every confidence" in me and that their trust in Lockheed would not be misplaced. I was 28 years old, quite a mature age, I thought. Courtlandt was 36.

Within a few days, on June 23, the Air Ministry gave Lockheed an order to build 200 airplanes of the model that became known as the Hudson, nicknamed "Old Boomerang" because it so often came back when very badly shot up. The contract also called for as many more than 200, up to a maximum of 250, as could be delivered by December 1939. This was

Johnson's mentor and organizational genius Courtlandt S. Gross who with brother Robert helped shape Lockheed's destiny.

the largest aircraft production order placed up to that time in the United States.

The British hadn't wanted us to discuss their equipment and plans over the transatlantic phone to the plant, so Hibbard and Gross and the others were quite surprised when we returned with an order for an airplane considerably different from the original design.

We came back on the German ship *Bremen* because its sailing time was much more convenient than the *Queen Mary's*. We wanted to return as soon as possible and get to work. Within 30 minutes after we had boarded, stowed our gear, and found the ship's bar, our cabins had been thoroughly searched. They knew who we were. Our plans were in the diplomatic pouch on board the *Queen Mary*.

We had burned all of our preliminary drawings in the fireplace at Mayfair Court, setting a fire in the chimney. So much carbon had collected within it over a great many years that it burned like coal. Chunks of it fell down along with flaming papers. Fortunately, the fireplace drew beautifully. We really cleaned it out. The fire scared hell out of us, though. We raced outside and saw the flames leaping up, then retreated quickly so we wouldn't call attention to it and our own involvement. We didn't want to have to pay for any damage caused. Luckily, there was none. The whole experience lasted only 15 or 20 minutes, but it seemed much longer at the time.

Taking on such a big production order took courage on the part of the Grosses. In 1936, the company had purchased more land in Burbank and the next year had expanded production, administration, and engineering facilities. Total floor space was 250,000 square feet, employment rose to 2,500, net working capital reached $650,000, and the company had about $334,000 in the bank.

After the British order was placed, a young bank vice president, Charles A. Barker, Jr., who had been following the growth of the company joined it as vice president for finance. He and Gross were able to raise $1,250,000 in short-term financing and, in 1939, Lockheed offered its first public stock valued at $3,000,000. That 250th Hudson was produced more than seven weeks ahead of schedule.

When the first three Hudsons were delivered, I returned to England with them for a flight-test and familiarization program, and to prove guaranteed performance. I was able to take Althea with me on the *Queen Mary*, and she loved life aboard ship. We danced every night and had a real vacation. In London, she loved to explore the city while I was working. Althea returned home when I went up to Martlesham Heath northeast of London for the flight-test assignment. The site was the British equivalent of the test center at Wright-Patterson Air Force Base in Ohio. I remained there about three months.

Lockheed test pilot Milo Burcham was with me. I particularly remember one event—the diving demonstration. We wanted clear weather for this and waited ten days or so before, finally, some holes appeared in the cloud overcast. We decided to chance it and took off when we had our calibrated barograph aboard and the plane loaded with the equivalent weight of arms. Almost immediately, the weather began to close in again.

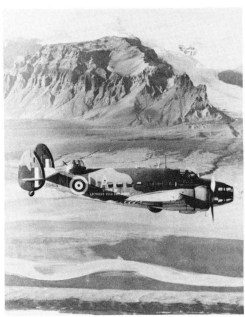

Althea Johnson, left, and Kelly, right, aboard the *Queen Mary* with the Lockheed team that sold the Hudson light bomber (left) to the British—the first in the company's series of successful maritime patrol aircraft.

"Let's try it anyway," Milo suggested.

After we'd climbed to altitude, we roared down with full power to our designed dive speed and leveled out. Low. So low that I remember distinctly flying by a cottage and seeing a woman looking out at us through the flowered curtains of her kitchen window.

"What pretty curtains," I thought.

We both were shaking slightly after in the eight minutes from takeoff to landing, but we stepped out as if we did that sort of thing every day.

The British drafted me into the Royal Air Force unofficially for the familiarization flights. They wouldn't take an American pilot, but assigned me—in an RAF blue flight suit—as flight engineer because I had to show them how to operate the engines for maximum range on fuel and demonstrate other operating procedures.

One of my early endeavors must have made the British wonder if I really knew what I was talking about. I wanted to show them what an excellent safety control we had on the landing gear—that it was not possible to retract it accidentally on the ground as you could with retractable gear in most other aircraft at the time. So one of the first things I did in the cockpit on ground inspection was to reach for the handle to show that it could not be raised, that it was held down by a solenoid. Of course, the handle came up. Fortunately, the weight of the gear was enough that the gear itself did not come up. Well, we re-rigged that and went ahead with the flight and familiarization. I had to prove all the performance figures we had guaranteed earlier.

It was in the course of these proving flights that I had my first dramatic encounter with the effectiveness of the English radar system.

The Hudson program was conducted under Commander "Red" Collins with whom I made many flights. He found the "iron mike"—the Sperry autopilot—the greatest of inventions. In the lead ship of a three-plane formation one day—with one Hudson fifty feet off our right wing and another fifty feet off our left wing—he put his plane on autopilot and proceeded to read the *London Times*. I hoped that the autopilot was good enough to fly us in formation for as long as this flight was going to last, but I went ahead with my work in the copilot's seat, leaning out the fuel flow to the engines to achieve maximum range.

We were aiming for a 2,200-mile flight to prove that the plane had that range without refueling and we had to fly all over Great Britain, Ireland, and the English Channel to cover that mileage. All was going well, but as we approached Scotland I saw dead ahead some huge black thunderheads we were going to fly right into. Commander Collins was still reading his newspaper.

"Look, Red," I interrupted, "I don't think we should tangle with that, do you?"

"Oh, my God, no!" He reacted immediately, reached up to disconnect the autopilot, and wheeled the airplane hard left. One Hudson went over us—I could see flames from its exhaust pipe—and the other passed underneath. Both went into the storm. We didn't, but had lost our formation.

With the aid of ground radar the three aircraft were rejoined.

We completed our range demonstration—and the *Times*—and flew back to base. But I had seen demonstrated the early English radar that was to serve so well later in the Battle of Britain. The ground crews had been able to locate all three aircraft in that stormy weather, guide us back into formation, and track us for the entire flight.

The RAF made good use of the Hudson, which performed as a fighter in the Battle of Dunkirk. In its primary role on anti-submarine patrol, it became the first airplane ever to capture a submarine. The U-boat had surfaced when spotted, and the Hudson kept its guns trained on the sub until a destroyer arrived.

Before the U.S. produced anti-submarine aircraft of its own, we actually had to borrow back some Hudsons. After this country entered the war, German submarine "Wolf Packs" began attacking our oil tankers within ten miles of the East Coast. Night after night they could be seen burning, and we had not a single anti-submarine warfare (ASW) airplane in the U.S. at the time. So we borrowed 19 Hudsons from the British and began to build some ASW planes for our own defense. Nearly 3,000 Hudsons were built by war's end for the British, Australia, and the U.S.

The Hudson, in fact, was the first in a long line of ASW aircraft, produced to this day, by Lockheed. The early ones, the Navy's PV-1 and PV-2 were derivatives of the Lodestar transport, a "stretched" Electra. New designs emerged later, for anti-submarine warfare is a very specialized, highly-sophisticated science.

On the PV-2, we did some pioneering work on "high-activity" propellers—out of necessity, a frequent reason. The aircraft's engines in the original design had so much power that we could not swing a propeller of the proper diameter to take advantage of it. We really needed a 17-foot propeller, which would have chopped about a foot into the fuselage! Starting with a new design, we put the engine nacelles far enough out on the wing to provide for the proper-diameter propeller. Ten feet, six inches was the largest diameter we could handle with that configuration.

So, to solve the immediate problem with the PV-2, I asked Hamilton Standard to reduce a 17-foot prop to 10 feet 6 inches to see how it would work. The prop was shorter but wider, grabbed a bigger bite of air while turning more slowly, and thereby avoided problems with air buildup at the tips.

Rapid technological development in propellers had begun in 1936 and '37. We were getting into variable pitch, full feathering, constant speed, and propeller reversing.

It was and is important for an engineer to keep up with advancing technology. Studying, fortunately, still held for me the same fascination that it had when I discovered the Carnegie library in Ishpeming. On one summer vacation in those early years, I reworked all of the problems in Fred Weick's classic book, *Aircraft Propeller Design*. On another vacation years later, I reworked every problem in Dr. Clyde E. Love's, *Differential and Integral Calculus*, which I had completed in college. I was determined not to lose my capability in mathematics. And I enjoyed both vacations.

For many years after I began my work at Lockheed, I would attend a Wednesday afternoon seminar conducted by the eminent scientists and engineers resident and visiting at Cal Tech. I also attended classes there, especially those taught by Dr. Clark B. Millikan, then head of the aeronautical department. It was my intention to earn a doctorate, and I completed all classroom work in the proper courses only to discover that there was a requirement then in my field for competency in technical German. There simply was not time for me to embark on that course of study.

When it became evident that we were going to need a wind-tunnel test capability of our own on a continuing basis, not dependent on scheduling of a rented facility, I was able to persuade the company to provide $360,000. The top officers always believed strongly in the need for research and backed our efforts.

From what experience I had in working with other tunnels and from work that had been done by the National Advisory Committee for Aeronautics (NACA), I undertook the aerodynamic design myself and assigned one of my best engineers, E. O. Richter, to draw plans for the structure. We put it out for bid and had the tunnel itself—the bare walls—built for $186,000 The rest of the money went for very expensive instrumentation, other construction work, and the model shop.

The result was a very good subsonic tunnel capable of testing to a simulated speed of 300 miles an hour. The test section housing the models was a rectangle twelve feet long by eight feet wide. The tunnel had a very useful constant-speed propeller system that was unique—and sometimes troublesome. In the Cal Tech tunnel, speed would change and have to be adjusted with each different angle of attack. With a simple electrically controlled drive, our tunnel would hold its speed well throughout the range of model position changes. We sold the design to six other companies for a modest $10,000.

That tunnel actually paid for itself on the first real test we put it to, because on our next big airplane design project, the P-38 fighter, we were to encounter a phenomenon about which very little was known—compressibility.

9

Into the Unknown

In the late 1930s, this country was awakening to a sense of its own unpreparedness for war, and for several years Lockheed had been at work secretly developing a new fighter for the Army Air Corps. When I got back to Burbank after introducing the Hudsons to service with the RAF, this became my first priority.

Specifications for the new fighter had been very clear—two liquid-cooled engines and a speed of 367 miles per hour. We advised the Air Corps that our design would fly faster than 400 miles per hour, a speed unequaled then. Lockheed received a contract for such a plane in 1937, with construction of the first beginning in July 1938. First flight of the XP-38—X for experimental, P for pursuit—was scheduled for early 1939.

It was considered a radically different design—even funny looking, some said. It wasn't to me. There was a reason for everything that went into it, a logical evolution. The shape took care of itself. In design, you are forced to develop unusual solutions to unusual problems.

For the new fighter, we were required to use the liquid-cooled Allison engine. This meant that we had to have a Prestone radiator. We had a long engine so we had to use a General Electric turbo supercharger. And we had a landing gear that had to retract into the nacelle. By the time we had strung all of that together we were almost back to where the tail should be. So, we faired it back another five feet and added the tail. It was a twin-engine airplane, and that produced the characteristic twin-tailed airplane that would go through 18 versions in all theaters of action in World War II, set some records, and make some design contributions.

The use of counter-rotating propellers on the P-38 was a new and important feature for fighters. It eliminated the torque effect, or pulling to one side.

With the first plane faster than 400 miles an hour, I knew we would be entering an unknown region of flight and possible trouble. It was the phenomenon of compressibility—the buildup of air ahead of the airplane at high speed. In 1937, in connection with our proposal, I had warned the Air Corps, ". . . as airplane speeds and altitudes (thinner air) increase, consideration must be given to the effect of compressibility."

When I first anticipated the trouble with compressibility, I went to the two best experts in the world on the subject, Dr. Von Kármán and Dr. Milliken at Cal Tech. I told them what I proposed to do in design, how we intended to compute performance, and our concern about stability and control.

"We don't know anything different. Go ahead," they agreed. Dr. Von Kármán had recently delivered a technical paper in Italy on the expectation of compressibility at higher speeds and predicted that shock waves would form on the wing, but he did not develop resultant effects on the airplane itself.

We encountered compressibility, but not immediately.

Our Air Corps project officer for the XP-38 was a pilot, a young first lieutenant, Benjamin S. Kelsey. He was excellent. In those days a project officer with that rank had more authority than many four-star generals do today. If we asked Ben for a decision, we got it—on the spot.

We had trouble before we got off the ground. We had trucked the airplane, under wraps for secrecy, to the Air Corps' March Field near Riverside, Calif., and reassembled it for first flight. The brakes had been received just the day before, because they had to be qualified first back at Wright Field in Dayton, Ohio, before we could install them. We had loaded the rudder with a 500-pound pedal force and the usual type of linkage to the brakes.

So, on a bright, sunny morning, Kelsey started up those wonderful-sounding Allison engines. He decided to make a high-speed taxi run. He got up to speed, then stepped on the brakes. No deceleration. He pushed and pushed—in fact he bent those pedals that we had tested to 500 pounds of pressure the night before. Fortunately, he was able to stop the airplane short of the end of the runway.

On disassembly, we found a small residue of grease from the oily rags in which the brake shoes had been packed. That didn't help the braking power at all. Well, we had located our trouble and were ready to fly. No committee inspections, no review. In those days, when we were ready to fly, we flew. Next day, first flight.

One of the design features of this airplane was the high-lift maneuvering flaps—it was the first fighter to have them. They were a fundamental part of the design and had to work.

Kelsey used a few degrees of flap setting, and the takeoff appeared to be splendid. Then the flap links broke, and all the flaps were sticking above the wing. Kelsey describes the experience:

"We developed wing flutter on takeoff. Looking out, we could see the flaps coming up above the trailing edge, so we retracted the flaps; the flutter stopped, but we had to come in on that first landing without any flaps—which was a little unusual."

We put a full-scale section of the wing with an actual flap and flap mechanism in our brand-new wind tunnel at Burbank, and that first series of tests proved the value of the tunnel. We found the solution quickly. We had a very fine streamlined aerodynamic gap between the flap and the wing, very sensitive to any small imperfections in air flow. So we just drilled holes in the fairing above the flap to let the basic structure control the air flow; and with air flow stabilized, the flap would not be buffeted. In the 10,000 airplanes that were produced, we never had flap flutter again.

When we finally ran into compressibility later in the airplane's development, many people thought it was a case of tail flutter because of the unorthodox appearance of the airplane. Kelsey was our staunch supporter in insisting that the problem was, instead, compressibility. Because the airplane was the first to get high enough and go fast enough to reach Mach numbers approaching the speed of sound—Mach 1—it was difficult to convince anyone that we had encountered this phenomenon.

When the first squadron of P-38s was delivered to Selfridge Field in Michigan, it was Col. Signa Gilkey who first encountered it in accelerated service testing, going through combat maneuvers and other performance tests that would subject the airplane to higher stresses than in normal flight. When Col. Cass Hough later reported exceeding Mach 1 in a dive over England, we knew that would have been impossible. The P-38 just could not have withstood that; it had to be an inaccurate airspeed reading because of instrument lag in the rapid change in altitude—speed of sound, of course, being different at different altitudes.

Several pilots were lost in the early days of testing. The aircraft would pitch nose down in an uncontrollable dive from which they could not always recover. It would happen at about Mach .67 to .80, building up rapidly once started. We tried all the usual methods to improve the elevators and made them so powerful that they pulled the tail off; our test pilot Ralph Virden crashed to his death.

Lockheed test pilot Marshall Headle, taking up the first of an initial production model, the YP-38, had proclaimed it the "easiest" plane he ever flew. The counter-rotating propellers produced no torque, or one-sided pull. But other test pilots—including Lockheed's Milo Burcham and Tony LeVier—encountered the compressibility phenomenon. "A giant . . . hand . . . sometimes shook it (the plane) out of the pilot's control," was how LeVier described it.

We had to find the solution in the wind tunnel; the aircraft was just too dangerous to fly.

Meeting with a committee at Wright Field in Dayton, I pleaded with NACA to let us put a model in its wind tunnel so

With a model of the P-38—one of the United States' most deadly fighter aircraft of the World War II era—engineers W. A. "Dick" Pulver, Johnson, Hall Hibbard, Joe Johnson, and James Gerschler.

that we could measure what forces were at work because there had been no high-speed tests of fighter or any other aircraft at that time. Our own tunnel could not achieve the required high speeds above 300 mph. NACA officials protested that every time they had approached such speeds in the tunnel, the model had thrashed around so violently they feared it would cause damage to their tunnel, a risk they did not want to take.

Lieutenant Kelsey went to Gen. H. H. "Hap" Arnold, then head of Army Air Forces, with the problem.

NACA got the message from the General.

"Put that airplane in the tunnel and run this test for Kelly. Find out what's wrong with my airplane. To hell with the tunnel. If it blows up, call me," was the gist of it.

We proved that the problem wasn't flutter, and we began to make some progress on a scientific approach to handling compressibility.

Although never solving the problem of compressibility, we

learned how to avoid it. In the NACA tunnel, we learned about pressure distribution on the wings, how effective was the tail, and what was causing the compressibility effect that pitched the nose down and resulted in such extreme buffeting.

On returning to Burbank, we decided that if we could not solve compressibility, we could discover a way to slow the airplane to a speed where the effect no longer was a factor. The answer was external dive flaps, or brakes. Put in the right place, they would cause the nose to come up out of a dive and stop buffeting. That right place was on the front wing spar. It also changed the pressure distribution on the wing's lower front so that the diving force was thoroughly counteracted.

It worked so well that if a pilot extended this flap and just let the wheel alone the airplane would pull itself out of a dive.

There were those who didn't agree that we should go to this expense and effort during a war. The airplane had been in service for some time before we confronted compressibility. Kelsey was one of the doubters.

"Let me fly that one with the compressibility flap on it, Kelly," he said on a visit to the plant. "I want to see what there is to this thing."

Kelsey certainly knew the XP-38. Shortly after its first flight he had flown it across country to Mitchell Field, Long Island— with stops at Amarillo and Dayton—at 420 miles an hour in seven hours and two minutes. Unfortunately, the engines lost power on the landing approach and in the emergency Kelsey pancaked on a nearby golf course. Kelsey was unhurt but the first XP-38 became a pile of rubble.

Kelsey took up the modified P-38 to prove to himself the need for the new compressibility dive brake. Flying from Burbank airport, he put the plane in a dive and soon encountered compressibility effects so extreme that he could not reach the dive flap switch. When the tail broke off and he was descending to low altitude at very high speed, Kelsey bailed out and broke a leg and sprained a wrist. He then became a believer in the dive flap.

Thousands of P-38s already were in service by that time, of

course, so in addition to incorporating the new flap into production design, we turned out dive flaps for the planes already overseas. For the U.S. Eighth Air Force in England, we sent 487 sets, along with aileron boosters and improvements for engine cooling at high altitude, that would make the P-38 the most maneuverable fighter in the world. It could have outclimbed, outrun, outmaneuvered, and outgunned any other airplane by a factor of two.

The modification kits were loaded aboard a military C-54 transport. As the plane was approaching the coast of Ireland, it was sighted and shot down by RAF fighters who mistook it for one of the German four-engine Condors that were threatening our convoys. The dive flaps never were put into service with the Eighth Air Force. But they were used in the Pacific later in the war.

We had another tragic loss although without the irony of being self-inflicted. A convoy with a ship-load of P-38s—more than 400—was headed for Murmansk and the Soviet Union when a submarine sank the ship. Those aircraft would have been our contribution to the Battle of Stalingrad. Both of these losses were at key periods of the war when the superior performance of the aircraft would have made a real difference.

Special highly-streamlined fuel tanks to extend range were introduced on the P-38. With 300-gallon tanks, the P-38 could fly more than 3,000 miles nonstop, unrefueled. Test pilot Milo Burcham said he made the flight on seven candy bars and one sandwich. The tanks were useful in combat during the last stages of the war in Europe and particularly in the Pacific. They also had other uses; for example, with the nose removable they became ambulance planes, capable of carrying a litter with a wounded soldier in an emergency.

The P-38 had acquired a troubled reputation because of compressibility, and some Air Corps pilots were reluctant to fly them. To counteract this, Lockheed test pilot Tony LeVier was assigned by General Doolittle to tour air bases in the US and in England to demonstrate the aircraft's capabilities. Tony, a daring pilot who knew exactly what he could get out of an airplane, did everything possible, including single-engine performance, in convincing demonstrations for the young military pilots.

Not only a top-notch fighter, the P-38 became very versatile—as camera plane, bomber-fighter, strafer, rocket-carrier. It went through 18 different versions, the last carrying a bomb load greater than the early B-17 Flying Fortress. It had excellent stall characteristics because of wing design and was particularly effective in combat against the Japanese Zero. The P-38 pilot could slow down to near-nothing airspeed, pull back on one engine, cartwheel without stalling, and reverse direction to face his adversary.

After the P-38, Lockheed built the XP-58. This was much larger than the P-38, almost the equivalent in weight and power of a four-engine plane. It carried a 75-mm cannon, with a cannoneer stationed behind the pilot.

Its role was to knock down big bombers, and it certainly could have if it hit the mark with that 75-mm cannon. We got the airspeed up to 450 mph, and the plane flew well. But it was very heavy and expensive. We built only two of them as experimental aircraft.

Because of the importance of compressibility as an aviation industry problem and the wide interest in it, I prepared a technical paper covering our own research and what we thought the solutions might be for presentation to the American Institute of Aeronautical Sciences. It was duly cleared by the War Department, and I presented it at a meeting in January 1943. Naturally there were many requests from other companies for copies, and I supplied them.

Then the paper was recalled and labeled "secret."

The agency charged with assisting, coordinating, and instituting this nation's aeronautical development did not want to acknowledge the work as industry-initiated. Later NACA did do some testing on its own but had contributed nothing to solving the problem of compressibility on the P-38 except allowing the use of its wind tunnel. And this only under orders from the Army Air Corps. The successor agency, National Aeronautics and Space Administration (NASA), by contrast, has been very aggressive and eager to assist and work with industry. I am happy to report that I enjoy excellent relations with NASA.

The matter of secrecy on compressibility became a moot point, for when the war was over and we were able to investigate, we found in German industry literature a great deal of information on compressibility, its effects, and how to avoid them. The Germans had handled it primarily with the swept wing, which they had been flying since the beginning of the Battle of Poland. Later claims to invention of the swept wing in this country are without foundation. By the end of 1943, the year I presented my paper, the Germans had all the aircraft types they were going to build, and the compressibility phenomenon was the talk of P-38 pilots everywhere.

In the course of World War II, horsepower for the P-38's engines had been increased from 1,000 to 1,750 per engine. Yet with all this great gain in power, we had been able to increase the speed only 17 miles an hour because of compressibility, the effects being felt first on the propeller long before they were encountered on the wing.

It became obvious that we would have to design better wings and tails, but that if we wanted higher performance we would have to get rid of the propeller.

10

The Big Time

AVIATION BEFORE WORLD WAR II had been for the pioneer, the daring record-seeker, the sportsman pilot, a few relatively wealthy travelers, government officials, and the military.

A new Lockheed transport, the company's first large one, would carry more people farther and faster and more safely than ever before, and economically enough to broaden the acceptance of flying as an alternative to train, ship, and automobile.

The Constellation was a tremendous challenge to Lockheed. It was our first attempt to enter the large-size transport field. Describing the company state of mind at the time, Hall Hibbard has said, "Up to that time we were sort of 'small-time guys,' but when we got to the Constellation we had to be 'big-time guys' . . . We had to be right and we had to be good."

Our commercial Model 14, so successful as the Hudson anti-submarine patrol bomber and the related Model 18 Lodestar—really a stretched Electra—were not large enough to compete in the expected post-war commercial air travel market.

Anticipating the future well before the war, we had worked on new designs, including Model 27, a canard, with horizontal stabilizer and control surfaces in front of the main supporting surfaces—or simply, with tail in front. We built a mockup but had the sense not to pursue this into production. The canard was impossible to make safe at high angles of attack—as the Russians later discovered with their supersonic TU-144 that crashed at the Paris Air Show in 1973.

Another was the Model 44 Excalibur, a very good "DC-4" in advance of the DC-4. It held considerable promise, and Pan American Airways expressed interest. Again, we built a mockup. Fortunately, we did not build it in prototype, as it would have been too small for competitive over-ocean service.

Then, in 1939, Howard Hughes as principal stockholder and Jack Frye as president of Transcontinental & Western Air, Inc., had asked Robert Gross if a transport could be designed to carry 20 sleeping passengers and 6,000 pounds of cargo across the US nonstop and at the highest possible cruising speeds. They suggested between 250 and 300 miles an hour at an altitude around 20,000 feet.

We abandoned our earlier studies and concentrated on the new airliner for TWA. What we proposed—Robert Gross, Hall Hibbard, and I—to Hughes at a meeting in his Muirfield Road residence in the elegant old Hancock Park section of Los Angeles was a larger airplane, capable of flying across the ocean and carrying many more people. We reasoned that it was economically unsound to carry only 20 sleeping passengers when we could accommodate more than 100 people in the same space with normal seating. Our design was capable of flying transatlantic with the Wright 3350 engine already in development for the military B-29 bomber. It was the world's largest air-cooled engine.

Few people not involved directly realize what a tremendous job it is to design, test, and build a new type of aircraft. And the larger the airplane, the more difficult are some of the problems. For example, the horizontal tail area of the Constellation was greater than that of the early Electra's entire wing.

The Constellation was first with many design features for passenger airliners. It was the first airplane to have complete power controls—that is, hydraulically "boosted." The basic principle of mechanically enhancing the human effort had been used in steamships, cars, and trucks, but the application to aircraft was much more complicated. Lockheed earlier had undertaken it as a long-range research project in anticipation of the greater control problems to come as aircraft performance increased. It was decided early to incorporate the device in Constellation design.

I had some difficulty in convincing Robert Gross of the advisability of adding this complexity to the airplane, since other aircraft manufacturers were ignoring it. Why did we need it? But I caught him one day when he had just parked his new Chevrolet in the company garage.

"Bob, you didn't really need power to steer that car, but it makes it a hell of a lot easier, doesn't it?" I never heard another word of dissent about power steering for aircraft.

The new airliner was to be faster, at 340 miles per hour top speed, than many World War II fighters. This soon was increased to 350 mph. Unlikely though it may seem, the Constellation transport used the same wing design as the P-38 fighter—larger, of course, and with an improved version of the Lockheed-Fowler flaps.

Its pressurized cabin allowed comfortable flight at 20,000 feet, above 90 percent of weather disturbances. It was the first airliner with this capability. Our early work on the XC-35 contributed importantly here. The airplane had excellent performance on just two of its four engines. And those powerful engines gave it an easy transcontinental and transatlantic nonstop range.

There were other innovations for transports introduced initially or in later development: integrally-stiffened wing structure, reversible propellers, turbo-compound engines, wingtip fuel tanks, and a detachable streamlined cargo pack carried under the fuselage.

That "Speedpak," the under-fuselage baggage carrier, was a very good concept and still is. It cost only 12 miles an hour in lost speed because of extra aerodynamic drag, and the passengers' baggage always was right there on landing. I wish we had done more with that; it never really caught on. Of course, airports were nowhere near so busy as they are today.

Six different wind tunnels were used in development of the original Constellation design. Most of the tests were conducted at the University of Washington and Lockheed's own tunnel, but supplementary tests were undertaken at Cal Tech and in NACA's high-speed tunnel, spinning tunnel, and 19-foot tunnel.

Engines were tested not only in ground runup but in flight, installed in a Vega Ventura. One in the ASW series of aircraft, this was produced by a Lockheed subsidiary. It gave us a flying test bench for the Constellation engine. Installation was based on several years of work by the Civil Aeronautics Board on fire prevention, warning devices, and fire-extinguishing methods. Despite these precautions, we later would have a temporary grounding of the airplane because of fire.

Development work on the Constellation led to establishment of a second major research and development facility at Lockheed—a laboratory for mechanical and structural testing of aircraft structure and systems. We had built a full-scale fuselage mockup of the airplane. Then, because of the complexity of hydraulically boosting the entire control system, we built a mockup of that alongside the cabin mockup.

With our limited space, this all was so crowded and so unprofessional looking that Messrs Gross and Hibbard took pity on us. "All right, we'll go for a research lab," Gross allowed. We had a start on the very extensive and sophisticated research and development facilities that exist at Lockheed today.

We built our new research lab next to the wind tunnel and made it large enough so that we could represent in full scale the entire control system of the Constellation from cockpit to tail. We could provide the equivalent of air loads on the control surfaces by the use of very heavy springs. This is how we developed what was the first power-boosted control system for any airplane.

The electrical system on the Constellation was another of those mocked up. This served a second purpose later when a TWA plane crashed on a crew training flight. The cockpit had filled with smoke from an electrical fire. We were able to reproduce conditions in the lab and also simulated the smoke conditions in actual flight—but we wore gas masks.

In the TWA tragedy, a short circuit in an electrical fitting had set fire to oil-soaked insulation and an open door then had allowed smoke to enter and blind the pilot and copilot. Our accident investigation resulted in some redesign with extra protection against the possibility of engine fire.

That "Day of Infamy," Sunday, December 7, 1941, put a hold on all commercial aircraft production. After Pearl Harbor, the Constellation project was stopped by military authorities who wanted Lockheed to concentrate on Hudson, P-38, and other war production. The Vega company, in pool with Boeing and Douglas, was to produce the B-17 bomber and abandon its original plans to build small civilian aircraft.

The Air Force, fortunately for the Constellation program, saw a need for military transport aircraft to carry large numbers of troops. The Constellation was "drafted." But its production was stalled 17 times by the military during the war because of the priority of other projects when the production people were needed elsewhere.

The Constellation made its first flight on January 9, 1943, in military olive drab paint, as the C-69. We had delayed the flight for two days because of very high winds—too high for a first flight with a large, new transport. The press corps—radio and newspaper reporters, press photographers, magazine writers, newsreel cameramen—would appear each morning only to be invited twice to adjourn to the air terminal Skyroom for breakfast and a wait while we hoped for the winds to subside, finally gave up and cancelled the flight. We were all happy when the third day dawned more gently.

The airplane made six successful test flights that day. Its accelerated service tests for the military at Wright Air Development Center, Ohio, set a record—170 flying hours completed in 30 days. The airplane also had the distinction of carrying in the cockpit Orville Wright on what would be his last flight.

At the end of World War II, Lockheed was in the enviable position of having a new, highly-advanced transport, thoroughly tested in military service and ready for commercial airline production. The first deliveries of an initial Model 049

Checking blueprints of a milestone undertaking—development of the Constellation transport—with Hall Hibbard. Test flight of the aircraft with Howard Hughes, below, was to prove a terrifying experience.

actually were conversions of Air Force C-69s already in work. It took only 90 days to turn out the first commercial model, which went to TWA in November 1945.

There were big plans to publicize introduction of this new transport in service with TWA. Howard Hughes himself wanted to be at the controls of what would be a recordbreaking, cross-country flight carrying press and Hollywood celebrities. He earlier had established a reputation as a pilot. In fact, he was awarded the Collier Trophy for an around-the-world record flight in 1938. He flew his Model 14 at an average speed of 206 miles an hour over a 15,000-mile route in 3 days, 19 hours, and 9 minutes. We had not worked with him on that venture, although it was with a Lockheed airplane. He had the extra fuel tanks installed on his own.

Hughes would have to be checked out in the new airplane before attempting the cross-country flight, of course. So, before it was delivered to TWA, Milo Burcham, Dick Stanton as flight engineer, and I took Hughes and Jack Frye on a demonstration and indoctrination flight. Frye was just observing, but Hughes was to learn how the plane performed and how best to handle it.

Our normal procedure in checking out a new pilot in an airplane was to go through the maneuvers carefully, then have the student follow through on the controls from the copilot seat.

We had just taken off from Burbank and were only a few thousand feet over the foothills behind the plant when Hughes said to Milo: "Why don't you show me how this thing stalls?"

So Milo lowered the flaps and gear, put on a moderate amount of power, pulled the airplane up, and stalled it. The Constellation had fine stall characteristics, not falling off, and recovering in genteel fashion.

Hughes turned to Milo and said, "Hell, that's no way to stall. Let me do it."

Milo turned the controls over to him. I was standing between them in the cockpit. Howard reached up, grabbed all four throttles and applied takeoff power with the flaps full down. The airplane was so lightly loaded it would practically fly on the slip stream alone. Hughes then proceeded to pull back the control all the way, as far as it would go, to stall the airplane.

Never before nor since have I seen an airspeed indicator read zero in the air. But that's the speed we reached—zero— with a big, four-engine airplane pointed 90 degrees to the horizon and almost no airflow over any of the surfaces except what the propellers were providing. Then the airplane fell forward enough to give us some momentum. Just inertia did it, not any aerodynamic control.

At that point, I was floating against the ceiling, yelling, "Up flaps! Up flaps!" I was afraid that we'd break the flaps, since we'd got into a very steep angle when we pitched down. Or that we'd break the tail off with very high flap loads.

Milo jerked the flaps up and got the airplane under control again with about 2,000 feet between us and the hills.

I was very much concerned with Howard's idea of how to stall a big transport.

We continued on our flight to Palmdale Airport, where we were going to practice takeoffs and landings. That whole desert area was mostly open country in those days and an ideal place for test flying.

Once on the runway, Milo and Howard exchanged seats. On takeoff from Burbank, Milo had shown Howard what the critical speeds were; so Howard now took the plane off. But he had great difficulty in keeping it on a straight course. He used so much thrust and developed so much torque that the plane kept angling closer and closer to the control tower. We circled the field without incident and came in for an acceptable landing. Then Howard decided to make additional flights, and on the next takeoff he came even closer to the control tower, with an even greater angle of yaw. He was not correcting adequately with the rudder. He made several more takeoffs and landings, each worse than the last. He was not getting any better at all, only worse. I was not only concerned for the safety of all aboard, but for the preservation of the airplane. It still belonged to us.

Jack Frye was sitting in the first row of passenger seats, and I went back to talk to him.

"Jack, this is getting damned dangerous," I said. "What should I do?"

"Do what you think is right, Kelly," he said. That was no great help; he didn't want to be the one to cross Hughes.

I returned to the cockpit. What I thought was the right thing to do was to stop this. And on the sixth takeoff, which was atrocious, the most dangerous of them all, I waited until we were clear of the tower and at pattern altitude, before I said: "Milo, take this thing home."

Hughes turned and looked at me as though I had stabbed him, then glanced at Milo.

I repeated, "Milo, take this thing home." There was no question about who was running the airplane program. Milo got in the pilot's seat, I took the copilot's seat, and we flew home. Hughes was livid with rage. I had given him the ultimate insult for a pilot, indicating essentially that he couldn't fly competently.

A small group was waiting for us at the factory to hear Hughes's glowing report on his first flight as pilot of the Constellation. That's not what they heard.

Robert Gross was furious with me. What did I mean, insulting our first—and best—customer? It was damned poor judgment, he said. Hibbard didn't tell me so forcefully that I'd made a mistake, because he always considered another person's feelings, but he definitely was unhappy and let me know it. Perhaps most angry of all was the company's publicity manager, Bert Holloway. He had a press flight scheduled that would result in national attention, headlines in newspapers across the country and in the aviation press around the world. Because, of course, the plane would set a speed record. Would Hughes follow through as planned? By that time, I didn't care what anyone else said. I went home and poured some White Horse and soda.

It was a frigid reception I received next day at the plant. But when I explained what the situation had been, that in my judgment I did the only thing I could to keep Hughes from crashing the plane, and then Hughes later agreed to spend a couple of days learning how to fly the plane as our pilot would demonstrate, the atmosphere thawed.

We offered a bonus to our flight crew to check Hughes out in the plane over the next weekend. Rudy Thoren, our chief flight test engineer, took my place. I never flew with Hughes again; it was mutually agreeable.

The only other time we ever thought it necessary to pay our flight crew a bonus to check out a customer's flight crew was in training a crew who would religiously release flight controls to bow toward Mecca at certain times of the day. It happened once on approach to landing!

On his next time in the airplane, Hughes changed his attitude considerably. He followed instructions carefully. He was the only pilot I ever knew, though, who could land one of our airplanes at cruising speed! He must have made 50 or 60 practice takeoffs and landings over that weekend. In fact, he was flying right up to takeoff time for the cross-country flight.

On the flight, as he was approaching Denver, Hughes encountered a big thunderstorm that had not been predicted. Instead of flying around or over it, and perhaps adding to the flight time, he plowed right through it. Unfortunately, the passengers had not been warned of turbulence and several not strapped in their seats were injured, though not seriously.

A record transcontinental crossing was set—Los Angeles to Washington, D.C., in an elapsed time of 6 hours, 57 minutes, 51 seconds.

From then on, the "Connie," as the plane soon came to be called, established records every time it first flew from point to point.

Except for the "penny-wise" policies of Hughes, TWA

would have had a monopoly on nonstop cross-country flight for some time, because no other airliner then in service could make the flight without stopping to refuel. But in winter, against maximum headwinds, the east-to-west flight especially would take longer than nine hours. A union rule required a change of crew after that period of time. Hughes would not double-crew the flights, although it would have paid off handsomely in competitive scheduling.

Hughes had exacted an agreement from Gross in signing the Constellation fleet purchase contract that Lockheed would not sell the airplane to any other airline until TWA had received 35 of them. Despite the fact that he thereby had prevented any other line from competing with him, he refused to take full advantage of the position if it meant double-crewing. The Constellation had no competition until Douglas brought out the DC-7 with the same turbo-compound engines.

The agreement with Hughes cost Lockheed dearly. It flawed our relations with American Airlines for years. It is interesting that AA doesn't fly a Lockheed-produced aircraft to this day, having opted for the DC-10 over the L-1011. In the commercial airplane business, everyone knows what everyone else is doing—although the information may not have been offered directly. American asked us if we could produce a new passenger transport for them, with specifications basically those of the Constellation.

Gross, Hibbard, and I met with C. R. Smith, then head of the line, Bill Littlewood, vice president and chief engineer— and a good engineer—and a few other American officials at the Ambassador Hotel in Los Angeles. We had to say that we could not build an airplane with that kind of performance although it went against the grain for all of us. And American Airlines knew damned well we were building such a plane.

American went to Douglas for the DC-6 and was responsible in large part for pushing continuing development of that plane until the DC-7 was introduced and then the jet-powered DC-8. American's decision in the '50s to buy hundreds of turboprop-powered Electras—it was named for the first Electra—from Lockheed effectively kept the company from being an early entrant in the jet transport field. The commanding lead that Lockheed had with the Constellation was lost.

Ironically, we may have explored the jet transport field too early. Before building the second Electra, we had invested $8 million in research and preliminary design of the Model L-199, a jet plane with four rear-mounted engines. Fuel consumption for the early engines was so high that it would require a huge airplane to fly across the ocean. Our design grew to 450,000 pounds for takeoff, and Gross decided that was just too big. Before turning thumbs down definitely, though, he retained a consulting firm for an opinion on the future of jet engines in commercial air transportation.

The report was discouraging because it forecast that operating time on the jet engine would never exceed 35 hours between overhauls. Now we think in terms of 10,000 hours.

Other orders for the Constellation followed TWA's, and eventually the airplane flew for most of the major airlines of the world, including even American Overseas Airlines. Its identifying triple-tail and graceful lines were recognized at airports internationally.

In its long production lifetime, the design was continuously improved and extended in performance and modified for a series of specific missions. The Constellation, and then the Super Constellation, appeared in more than 20 successively advanced airline versions, cargo models, and a series of early warning, patrol, and other specialized service-types for both U.S. Air Force and Navy. Some of the commercial liners on long-range flights did have berths for sleeping passengers. The fuselage was stretched and the wings extended.

The last of the airliners, the Model 1649, still is remembered by air travelers for its luxurious interior. The planes had reclining seats with retractable footrests in standard cabin configurations. Had the turboprop engines for which this Super Constellation was designed been available, the series surely would have had an even longer life. But it could not compete with the coming jets. The last commercial Constellation was produced in 1959.

There was a myth circulating for some years that Howard Hughes had designed the Lockheed Constellation. It was not discouraged by Howard, and certainly was not true. His specifications had consisted of half a page of notes on the size, range, and carrying capacity he wanted. It was not without some encouragement from us—I did not appreciate someone else's taking credit for our work—that eventually both Hughes and Frye acknowledged the misconception in November of 1941. They offered to publish advertisements, but Robert Gross was satisfied that their letter stated: ". . . to correct an impression . . . prevalent in the aircraft industry . . . the Constellation . . . airplane was designed, engineered and built by Lockheed."

Hughes used to keep at least one Constellation parked on our flight line—he had one of just about every type of plane stashed away somewhere; and he would phone his favorite flight test engineer at Lockheed, Jack Real (now head of Hughes Helicopters), in the early morning hours about once a month, wanting to come over, climb into the cockpit, run up the engines, and just sit there awhile. Real would join him.

The personal eccentricities that later were to become obsessions and make a tragedy of Hughes' life had not yet manifested themselves—at least, not to us. Hughes and Real became good friends.

While Hughes and I never again flew together, I heard from him directly during the period when he was developing his wooden Flying Boat, now a tourist attraction alongside the *Queen Mary* at Long Beach harbor in California.

The project actually had been proposed by Henry Kaiser, who chose wooden construction because material was plentiful while aluminum was in critical supply for war production. Hughes embraced it enthusiastically; he had the plant and people available. He took to telephoning me—I remember specifically one 8:00 a.m. call on a Sunday, because it was unusual. Hughes generally telephoned only late at night or very early in the morning. He said, "Kelly, we're going to build a nacelle like this. . . . What do you think of it?" I made my comments, as I usually did, and after about two hours managed to escape and go about my own activities.

He did this on several occasions, on different subjects. Each time, I discovered, he then would call Gene Root, chief aerodynamicist at Douglas at that time, and say, "Gene, Kelly says. . . . What do you think?"

Then he would phone George Shairer at Boeing with "George, Kelly said . . . and Gene said . . . What do you think?" So he had a three-way consultation.

Hughes had an excellent design team on that boat, and also on his FX-11. The FX-11 he later crashed on its first flight at night over the city of Beverly Hills and was hurt quite seriously. The plane was a twin-boomed fighter-reconnaissance plane, as was the P-38. To this day, Hughes' claim that the P-38 design was based on his FX-11 will appear in print from time to time. The FX-11 first flew in 1946. By the end of 1944, Lockheed already had built and delivered for military service some 10,000 P-38s!

The Flying Boat was a sleek design, about as good as the state of the art at the time would allow but far from being capable of carrying 750 people across the ocean efficiently and economically. It was heavier in wood than it would have been in metal, and a lot of its potential payload disappeared right there.

Hughes was determined to fly that boat to say that it had been done. Its first and only flight, with Howard at the controls, was a mismatch of common sense and responsibility. He had aboard about 32 people, not crew members but newspaper people and other guests who thought they were going for a high-speed taxi test. As he taxied down the harbor, he lifted the boat about 100 to 150 feet, and flew about a mile. If the airplane had gone out of control it would have been a tragedy. These people had not intended to go flying, particularly on a first flight.

While this determination to prove himself in aviation sometimes led to seemingly heedless action, it also spurred him to seek a forward-looking passenger transport for his airline—a decision from which air travelers all over the world would benefit in safer, faster, and more enjoyable flying. Hughes deserves credit for that. And his commitment to purchase the advanced airliner that became the Connie put Lockheed into the big time of commercial aviation.

What the Constellation was to air passenger traffic, the C-130 Hercules later was to the air-cargo business. Introduced as a military plane, it was the first design specifically for that purpose. During World War II and after, bombers or troop carriers had been converted for cargo. It was the first transport designed from the drawing board up to take advantage of new turboprop engines—jet engines driving propellers. They promised speeds of 300 to 500 miles per hour at altitudes to 45,000 feet.

With the engines and its special design, the C-130 was a major development in cargo aircraft. It would fly higher and faster and more economically than existing military transports and was tremendously versatile.

The fuselage was so low—only 45 inches from the ground—that loading was easy under any conditions. A section of the aft fuselage dropped down to become a loading ramp. The airplane converted easily and quickly from personnel carrier to hospital ship, from flying a load of heavy machinery to dropping paratroopers.

It was designed to land and take off from short and rough runways. It even operated from a carrier in a demonstration of performance. The plane was designed in what was to be known as Lockheed's "Skunk Works" but assigned to Lockheed's Georgia company for production. Later commercial versions were developed there. The plane has been a workhorse around the world. Many of its most effective features later were adapted to the much larger C-5A, designed at the Lockheed-Georgia Company.

11

The Jet Age— and the First "Skunk Works"

Since 1940, LOCKHEED HAD BEEN AFTER the Air Corps to build a pure-jet-powered airplane.

The potential of jet power had interested us for some time, particularly as we encountered the effects of compressibility on the P-38 fighter, first on the propeller, then on the wings. We decided there had to be a major change in the power-plant configurations for our fighter aircraft. During World War II, we had been able almost to double the power of the P-38, yet succeeded in increasing the speed only by about 17 miles per hour.

Independently, we developed a preliminary design for an airplane that would approach Mach 1—the speed of sound. It would be powered by a jet engine designed by a Lockheed consultant, Nate Price, a designer of great vision and knowledge of thermodynamics, materials, and mechanical design. We proposed to the Air Corps that Lockheed be permitted to build a prototype. The response was negative. We were told to devote our energies to solving the problems with the P-38 and other immediate wartime projects. In retrospect, of course, that was short-sighted.

But in 1941, when the British installed Frank Whittle's jet engine in one of their small fighters, the Gloster Meteor, and demonstrated the speed potential, that attitude changed. The Air Corps commissioned use of the engine in the Bell P-59, originally designed as a propeller-driven airplane. When the jet-powered version flew in 1943, the performance was hardly better than that of the piston-powered P-38 and P-51.

Again we proposed to build an airframe and jet engine in a very short span of time. This time the Air Development Center was receptive.

The Germans by this time already had a number of jet-powered Me-262s in combat, and these planes were much faster than anything we had. They were well into the jet age while we were just starting. The Me-262 was a very good airplane, designed by Willy Messerschmitt, whose talent I respected.

One argument against our pursuing jet power had been that the airplanes were fuel-hungry and lacked range. The Germans didn't have very far to go to reach England because they could take off from occupied territory in Holland and France, while the U.S. had to consider the large expanse of territory to be covered in Europe and in the Pacific. From Great Britain, we had to get to Berlin to be effective.

Within a week of hearing from Wright Field, I was back again in Dayton to present our design.

"We'll give you a contract for the airplane, Kelly, and for Nate's engine as well," said Gen. Frank Carrol, commanding officer of Wright Field. "But you'll have to use the British engine in the first airplane because we need it—and all the jet fighters you can build—as soon as possible to use against the Me-262. Your new engine couldn't possibly be ready for service in time."

Since I had promised to build a jet airplane within 180 days, I asked, "When will we get a contract? When will the time start?"

"You will have a Letter of Intent this afternoon by 1:30 p.m." he replied. "There is a plane leaving Dayton for Burbank at two o'clock. Your time starts then."

And it did. The date was June 8, 1943. Gen. H. H. "Hap" Arnold, himself, had approved the contract.

Back in Burbank, I knew I was in for a rough time. Lockheed already was producing 17 P-38s, four B-17s, and enough Hudsons, Lodestars, and PV-1s to total 28 airplanes each day. That was with three shifts a day, six days a week, and some work on Sundays. There were no spare engineers. There was no spare machinery and no available space.

When I showed Robert Gross a contract for our first jet fighter, he expressed some doubt that much would come of it. But he and Hibbard always were open to new ideas and backed me in many critical times.

"You brought this on yourself, Kelly," Gross stated. "Go ahead and do it. But you've got to rake up your own engineering department and your own production people and figure out where to put this project."

For some time I had been pestering Gross and Hibbard to let me set up an experimental department where the designers and shop artisans could work together closely in development of airplanes without the delays and complications of intermediate departments to handle administration, purchasing, and all the other support functions. I wanted a direct relationship between design engineer and mechanic and manufacturing. I decided to handle this new project just that way.

Absolutely the only place we could think to put the new work was adjacent to the wind tunnel. We already had a shop there to build the tunnel models, so this was the beginning of our machine shop. To get more tools, we had to buy out a small local machine shop. We found a lot of Wright engine boxes left over from deliveries for the Hudson bomber. They were just taking up space in the storage area and were made of good, heavy wood. We cleared the space and used the boxes to build the walls of our production area. For the roof, we rented a circus tent.

Somehow I got together 23 engineers, counting myself. I simply stole them from around the factory. I wanted people whose work I knew. Assistant project engineers were W. P. Ralston and Don Palmer, my good friend from college who by that time had joined Lockheed. The single-engine transport on which Vultee had staked its future—and on which Don was working—could not compete with Lockheed's Electra, Boeing's 247, and Douglas's DC-2—all twin-engine models. Art Viereck was recruited to head the shop. We had our own purchasing department and every function we needed to operate independently of the main plant. This became the first "Skunk Works."

How did it get that name? I'm not sure. But in the strict secrecy of wartime, and simply for efficiency and to avoid distractions, we allowed no one who wasn't working on the project to wander in and out. The legend goes that one of our engineers—and I'd guess it was Irv Culver, a brilliant designer—was asked, "What the heck is Kelly doing in there?"

"Oh, he's stirring up some kind of brew," was the answer.

This brought to mind Al Capp's popular comic strip of that day, "'Lil Abner," and the hairy Indian who regularly stirred up a big brew, throwing in skunks, old shoes, and other likely material to make his "kickapoo joy juice." Thus the Skunk Works was born and named.

When the Air Corps decided to move, it moved fast. Nine days after our go-ahead, we had a mockup conference at the Skunk Works with Col. M. S. Roth and Maj. Ralph Swofford, who became our project liaison officer. There were only six people from the military involved and two or three of us from Lockheed. We had approval to proceed that night. Six days later we had our government furnished equipment—guns, radio, wheels and tires, etc. At every stage of the work, we had excellent cooperation from Wright Field and the officers involved with the project. The job could not have been completed on such a tight schedule without it.

Pressures on everyone in our plant became intense as we counted off the contract days on a big scoreboard calendar. We had scheduled the work on the basis of a 10-hour day, six-day week. No one worked on Sunday. We had to enforce that rule because even with it the sickness rate during the last few weeks went as high as 50 percent daily. It was midwinter, and we had poor facilities in which to work and almost no heating. We could not afford to lose even one of our small supervisory staff to illness.

The airframe was never a problem, but the engine was. It didn't arrive until seven days before we were ready to fly—we'd been working with a wooden mockup to design its installation. It didn't help that the De Haviland engine expert sent over to help us had to spend time in a local jail. Guy Bristow had arrived secretly with the engine on an Air Corps plane without the usual travel papers. He made the mistake of jaywalking on Hollywood Boulevard and when the police cited him and discovered he had no draft card and wasn't even a citizen, they detained him. When we discovered his plight, we had to call on the Air Corps to negotiate his release.

During a final engine run-up before the first flight scheduled for the next day, there was a tremendous bang. I was standing between the two engine ducts, watching the operation, and I almost lost my pants down the intake. The ducts had collapsed, pieces of metal had gone into the engine, and the compressor housing had cracked. The engine was beyond repair—the only engine we had. We had to wait for another.

Finally, on day 143, the plane was accepted by the Army Air Corps and ready for flight. We had beaten the schedule of 180 days. On the morning of January 8, 1944, the XP-80 flew for the first time. We had named the plane "Lulu-Belle"—we always nicknamed our "firsts." Milo Burcham was at the controls at Muroc Dry Lake, now Edwards Air Force Base. It was the first American fighter to exceed 500 miles an hour—502 mph was top speed.

By this time the Air Corps had General Electric working on a bigger and more powerful version of the Whittle engine. Development was moving fast. Having proven our design with the first XP-80, we were asked to do the job over again with an airframe about 80 percent larger to accommodate the new GE engine. We were given a contract to build two such airplanes.

In 132 days, we built the first of these YP-80As. We gave it a coat of slick lacquer paint, off-white in color, so its nickname was the "Gray Ghost." It had heavier armament, carried more fuel with extra wingtip fuel tanks, and added about 80 miles an hour to the speed of our first jet.

It was the forerunner of the P-80, later redesignated F-80, and its successor T-33 two-place trainer, and the derivative F-94 in A, B, and C versions. When the Air Corps ordered the plane into production, more than 6,000 were built in all. It became this country's first tactical jet fighter. Robert Gross named it the "Shooting Star" in the Lockheed tradition of naming aircraft after stellar bodies.

Development was not without problems, of course. It was a complete new world of flying and testing.

The first problem we encountered had to do with compressibility again. But it was not the same as with the P-38. Here it came in shock waves jumping back and forth across the hingeline of the wing and aileron. Shock pressure changes would drive the ailerons and make them "buzz" at a very high frequency. We had known that we would encounter compressibility in some aspect with the airplane, and it was experienced at speeds of Mach .8 to .85. The pilot would get some warning. He would experience a little control stick shaking and feel the aileron "buzz" in the cockpit.

This effect we were able to overcome with hydraulic dampers acting as shock absorbers located on the aileron and attached firmly with no slop, or freedom, of movement.

Again, we used wind tunnels to find out what degree of safety we had. We fluttered a full-scale F-80 wing for more than 100 hours until it finally tore loose at the hinges and the ailerons flew down the tunnel. The NACA flew the airplane and exceeded that mark by flying to Mach .86. My hat is off to that pilot, because the test really made a shambles of the ailerons. But this top speed never became a limiting factor since the plane had more than enough speed for the competition.

A problem developed with the new engine. Test pilot Tony LeVier was flying F-80 No. 3 on a high-speed run at altitudes of 15,000 to 20,000 feet when suddenly there was a booming sound so loud that we could hear it on the ground. As we looked up, the airplane disintegrated and a parachute blossomed.

As Tony explained, "I was sitting there all fat and happy,

when the airplane suddenly flipped over on its back, the wings broke off, and I was sitting out in space." He had to release the cockpit canopy, of course, to get out. It took Tony about ten minutes of excited recollection to tell us this story once he had landed.

But a local farmer who had witnessed the same thing from the ground put it this way in less than ten seconds: "I looked up, there was a boom, and then a parachute."

When we had collected all the aircraft and engine pieces. we found that the jet turbine disc had broken into three parts, slicing through the fuselage. The reason for the failure is a problem to this day. We do not have in this country a machine big enough to forge these large metal aircraft parts; they must be welded—and that always presents a possible failure point. The largest machines we have are the few we captured from the Germans after the war.

On the F-80, the engine had a weld in the main shaft that came off the big turbine disc, because the biggest press in the country at that time could not forge it—hub and disc—together. Six F-80s were lost for that reason.

We changed the design of the hub as much as we could, then spun each turbine with its shaft in a vacuum to speeds faster than' operational. That always worried me—it seemed somewhat like seeing how far you could hang out a window of a ten-story building. The test procedure itself could have started a crack, which might then have been missed in X-ray, and the plane flown with this fault. However, this did not happen and the solution worked.

With the F-80, the Air Corps was able to develop tactics for flying against the German jets—both with bombers in defense and fighters in attack, not only with single aircraft but in formation. We spent several weeks flying with the military at Edwards. All the various U.S. fighters were there—P-38s, P-39s, P-47s, and P-51s—to be evaluated against the jet as escorts for our bombers. They were armed with gun cameras to record their "kills." So were the B-17s and B-24s.

I spent more than five hours each day, at 25,000 feet—

wearing tennis shoes, shorts, and a parachute—riding "piggyback" in a modified P-38 with Tony LeVier, watching them try to gun down the jet. We got into some very fancy maneuvers; all of us spun in trying to turn into the F-80. I must confess I enjoyed it; the P-38 was a good airplane for spins, easy to pull out.

"I got him! I got him!" the gunners would exclaim. But back on the ground, their film never showed a hit on the F-80.

The F-80 would make head-on passes at the bomber formation, roll over, and pass underneath inverted. Lateral and tail attacks were practiced. The fighters were flying defense. The air was crowded with airplanes.

These mock flights were very valuable. We discovered that while it was impossible to stop a frontal attack by the jet, it didn't matter much. At a closing speed of more than 700 miles an hour the probability of an Me-262 pilot's scoring a hit in the few seconds' firing time he had while avoiding collision with the bombers and fighters was so remote that we decided not to worry about it. The main concern turned out to be attack from the rear quadrant, where the German jets could overtake our aircraft, fly at any matching speed, and have considerably more time to aim and fire. To counter this, the Air Corps developed the tactic of having the fighters watch the rear. These exercises with the F-80 saved the Eighth Air Force from having to discover in combat that characteristic of the German jet fighter.

One night test that provided valuable information also resulted in a tragic loss. An F-80 took off with a B-25 as observer to determine whether the jet left a telltale glowing exhaust trail. It didn't, and the two planes collided in the darkness. Lockheed test pilot Ernie Claypool and the military pilot were killed.

Before World War II ended, four F-80s were sent to the European theater. They were there for indoctrination flights only, on patrol from England and Italy. We didn't want them shot down where the Germans might capture them. The Air Corps needed familiarity with the airplane, with its high fuel consumption, high speed, and operating characteristics in varying weather conditions. We lost one of these planes and its pilot. Very skilled and experienced, Maj. Frank Barsodi completing a high-speed, low-altitude pass apparently decided he was going too fast for a landing and pulled back on the engine throttle in flight. The result was more pressure from ram air on the outside of the tailpipe than from within, and the tailpipe collapsed inward. At least, that is what I think happened. This was an unforeseen occurrence because its pipe had been designed to withstand inside pressure. The modification here was to vent the pipe so never again could the pressure become higher on the outside than inside.

Development of the F-80 was to cost yet another life—that of our long-time Lockheed test pilot Milo Burcham. In the first production model, as distinguished from the experimental models trucked to Edwards for flight test, Burcham took off from the east-west runway at what then was Lockheed Air Terminal, now Burbank–Glendale–Pasadena Airport. In those days there was a lot of open space where a pilot could pancake in if necessary. But there also was an open gravel pit near the runway. Barely off the ground at 200 feet, Milo lost all power. The drive to the fuel pump shaft had sheared the spline. He couldn't avoid the gravel pit, and the plane exploded on impact. The design result was an improved spline and pump and installation of an emergency fuel system.

Ever since then, Lockheed aircraft always have had standby fuel systems, either double fuel pumps on the engine or a fuel pump on the engine and another with electric drive. Some other jets do not have this. Redundancy in systems since then became a mania with me. With everything we build, we make sure that we can relight, restart, and keep flying if the main engine pump fails. Milo's death contributed to future safety for others.

The war in Europe ended before the F-80 could be proved in combat there. But development, testing, and production continued.

When the Air Corps team went to Germany after the war to inspect military capabilities, we at Lockheed were invited. I stayed home to work on the F-80 development, but my assistant, Ward Beman, made the trip and picked up a great deal of information.

We found that the Germans had been flying the only axial flow jet engine in the world, fundamentally more efficient than the centrifugal compressors of the British jets because it was of simpler design. The flow went straight into the inlet and progressed in a straight line through the engine and out the exhaust. In centrifugal flow, the air goes in two sides of a rotor, flows perpendicular to the flight path of the airplane, enters the burner cans, then goes through the rest of the machine; so that it changes flow 90 degrees at least twice. But we and the British had a great deal of experience and knowledge of centrifugal compressors from use with steam power plants so that had seemed a surer and safer development.

Postwar, the Air Corps dramatically demonstrated the capability of its new air-defense weapon. In January 1946, three F-80s departed the West Coast on the same day for cross-country flight—two with a refueling stop, one nonstop.

Col. William Councill took his F-80 from Long Beach to La Guardia Airport in New York in the record time of 4 hours, 13 minutes, covering the 2,470 miles at an average 584 miles an hour. The feat was widely heralded in the nation's press. Councill's plane had carried 300-gallon wingtip fuel tanks dropped over Kansas farmland when empty. We heard later that the tanks were cut in half by farmers and used as feed bins. Refueling stops—at Topeka, Kan.—for the other two planes were completed in just a few minutes, so there was very little difference between the one-stop and non-stop flight times.

There was an amusing aspect to the record-setting takeoff. The official FAA timer came over to our chief flight test engineer, Rudy Thoren, pulled out a dollar watch, and asked Thoren the time! Of course, the quality of his watch didn't really matter. All he had to do was establish the starting time in any manner. Someone at the finish would record landing time. But the official image suffered somewhat.

More records fell. In 1947 when the Air Force became a separate service, Col. Albert Boyd averaged 623.8 miles per

Congratulating Tony LeVier following a successful test flight during development of the F-80, the nation's first tactical jet fighter. Below, even by today's standards the Shooting Star looks sleek and swift in this earlier photo.

hour over a measured course for a world speed record. Incidentally, jet fuel in those days, more like kerosene than gasoline, cost about 13 cents per gallon, not the $1.50 it costs today in the '80s.

The F-80 had proved so easy to handle that it took an effort to interest the Air Force in a training version. We made a piggyback version first—that is, with an enlarged cockpit so an observer could ride behind the pilot. Then we took a plane off the production line—had it disappear from the books temporarily—and converted it to a two-man model for demonstration purposes to convince the military of the need for a two-place version. Once convinced, the Air Force bought thousands of them, as the T-33. The Navy bought a trainer version, too, the TV-1.

Sometimes it's awfully difficult to convince the customer of what we think he needs, and sometimes we don't succeed. One such proposal was our Universal Flight Trainer.

In 1954 we suggested equipping the T-33 with "black boxes" to simulate performance of any other kind of aircraft—long before we had such flight simulators in the laboratory. But I could not sell the idea. Airplanes have been used in this manner since—both the JetStar and the F-104 Starfighter have been instrumented to simulate various phases of Space Shuttle performance, for example.

The current "new generation trainer," I believe, will have some of this capability. Such a "universal trainer" still would be practical and productive in research and development.

The F-80s in the Air Force were able to prove themselves in combat when North Korea invaded South Korea in 1950. In history's first jet battle, an F-80 shot down a Russian-built MiG-15. And that dual fuel pump that resulted from Milo's death brought a lot of pilots home from combat in the war zone. Jet combat in that theater provided other lessons we would use in aircraft development. We knew we had enough to keep us busy for some time.

12

Lessons from Korea

WHEN TEST PILOT TONY LEVIER FIRST SAW the F-104 Starfighter, he asked, "Where are the wings?"

It is true that the wings were short, straight, and almost razor thin, but they carried quite a load. They were the result of tests on some 50 wing models fired on instrumented rockets attaining speeds of 1,500 miles an hour.

The "Missile With a Man in It" resulted from my tour of the Korean battlefields in 1952. Lt. Gen. Benjamin Chidlaw, chief of Materiel Command, and other Air Force officers wanted to discover—and wanted the aircraft designers to know—how our aircraft were performing and what our combat pilots needed in confrontation with the enemy. This was the first war in which both sides had jet aircraft. The North Koreans—in effect, Chinese—had the MiG-15. The South Koreans—in effect, USA—had the F-80, F-84, and later the F-86.

It was an education to see our pilots operating from forward bases like Taigu, their aircraft so heavily laden on takeoff with wingtip fuel tanks to fly into enemy territory that the tanks would scrape the runway. The runway was made of steel planking and there was danger of fire, but they did it day after day.

We interviewed these pilots as they climbed out of their aircraft on return from missions. What did they want in a fighter? It was unanimous. They wanted speed and altitude. In combat at fighting altitude the two sides were about equal in speed. We were much more maneuverable with power controls, and our pilots had the advantage of the latest gunsights. Overall, our margin of victory was something like ten to one.

But our pilots were insulted constantly by "High-altitude Charlie," sitting at 50,000 feet and directing the MiGs in Chinese or Russian.

"Don't worry about the American airplanes. Your airplane will take care of you. Just come up here. They can't get up to you," our pilots heard constantly.

On our tour of the Korean battlefronts, we covered more than 23,000 miles and visited 15 air bases, flying in an Air Force Constellation.

On night flights, my civilian companion, Lee Atwood, vice president of engineering for North American Aviation, and I slept on a piece of plywood placed on the floor. I could detect vibrations from number three engine coming through a rear spar. The engine ran rougher and rougher but held together as we flew to Hawaii, then Wake, up to Japan, around the Korean air bases, down to Okinawa, and on to Manila. Just as we attained cruising altitude on takeoff from Manila, there was a big bang and number three was dead. We dumped fuel and returned. The Air Force had carried enough spare parts for a minor overhaul, so within 12 hours we were on our way again.

When we returned to the U.S., I set out to design a jet fighter that would fly higher and faster than anything flying anywhere. There was no formal requirement from the Air Force yet for such a plane, but that was cleared up very quickly with a visit to the Pentagon. When I showed my proposal to Gen. Don Putt and Don Yates and Col. Bruce Holloway, they were very receptive. The only obstacle to a contract was the absence of a document specifying what the next fighter should be.

"Well, if there isn't a requirement, I'm going to make one," Colonel Holloway decided. "Stick around, Kelly. Come back in a couple of hours."

He then wrote a page and a quarter on what the Air Force required for its next fighting plane. It should be lightweight, capable of certain performance at sea level and high altitude, and carry specific armament, radio, and other instrumentation.

In that short span, he had obtained the approval of all the appropriate generals.

"Here are your requirements," he told me. Generals Putt and Yates agreed. "Go see what you can do with this."

The result was the F-104 Starfighter, which would become the SuperStarfighter, and evolve through eight different models and production in the U.S. and six other countries of the Free World over a period of more than 25 years. This was to be the largest international industrial collaboration in the world. The airplane eventually would be flown by 14 allied air forces as well as the USAF. A two-place version also was built, not only for training but for tactical use.

The F-104 became the first operational airplane to fly twice the speed of sound in level flight.

Development of those short, thin wings utilized a new test technique—out of necessity. We didn't have a wind tunnel capable of going to Mach 2, the speed for which we were designing the airplane, so I visited Lt. Gen. Earle E. "Pat" Partridge, in charge of the Air Research and Development Command in Baltimore, and described the problems we were having in testing the thin wing as well as the tail and air intake ducts for the new fighter.

"What can I do for you?" he asked.

"Well, if we had a bunch of five-inch rockets, we could put the wing models on the rockets. If we shot enough of them we could find out how to make a wing with the thickness we want—about twice that of a razor blade. We can see if it flutters or not in supersonic flight."

Immediately, General Partridge sent a message to Korea, "Stop shooting rockets for one morning and send them all to Kelly."

Two weeks later when I returned to Burbank, the Skunk Works was in an uproar. There were about 460 rockets on hand and no one knew what to do with them. It wasn't a good idea to store them in the middle of Burbank, and, of course, we couldn't fire them from the city. We moved them all to Edwards Air Force Base, where there was plenty of open country within the security of the test base boundaries.

We fitted the rockets with automatic cameras and tele-metering equipment to radio data to ground observation stations. We put wing after wing on instrumented rockets and fired them from the desert base. Wings with different stiffness, different shapes, different designs, at speeds up to 1,500 miles an hour. There had been some question within the industry about the advisability of making a wing as thin as we proposed. Not only did the final design prove solid in test flight and in service, but we later hung wing tip fuel tanks on it, a great deal of armament, and even added A-bomb carrying capability.

The unusual "flying" tail configuration also was tested initially on these rocket "wind tunnels."

With the first F-104, we faced again the problem of starting work on an airplane before there was an engine for it. So, like the F-80, the F-104 flew with one engine in the first two prototypes before acquiring the engine for which it really was designed—the General Electric J-79.

The first airplane was built and flown in a year and a day. Tony LeVier took off from Edwards AFB on February 28, 1954. I like to schedule first flights on my birthday when I can, but I missed this one by a day.

That new GE engine gave us trouble early in the flight program. The problem was in the control to the afterburner—which injects extra fuel into the hot exhaust gas, burning it to raise temperatures and speed flow out the tailpipe. It can double the thrust when an extra burst of power is needed for takeoff or in flight.

On this engine the afterburner eyelids had a nasty habit of opening unexpectedly under certain conditions, resulting in an almost total loss of thrust. We built our own engine test tunnel specifically to run tests ourselves and try to speed redesign of the engine.

Before it was corrected in engine design, the problem cost seven planes and seven pilots. That still sticks in my craw. The fix was very slow in coming.

The gun caused problems, too, in early firing tests. The

Lockheed test pilot "Fish" Salmon with F-104 Starfighter.

flights required all the expertise of veteran pilots LeVier and Herman "Fish" Salmon—who flew the first production model of the F-104—as distinguished from the first experimental model flown by LeVier.

For LeVier it meant making the first dead-stick landing in the airplane. He was airborne on the first cannon-firing test at supersonic speed when the cannon exploded and blew a hole in the fuel cell. With a cockpit full of smoke, Tony's first thought was to bail out. But 50 miles from home and at high altitude, he looked down and thought better of it.

"Oh, my God, I'll freeze to death before I get to the ground," he decided.

"I stayed in," he recalls, "and headed back to base. As I neared the base, the engine quit. And at the very last, just before my flare for landing, I discovered that my wing flaps wouldn't work properly. Again I wanted to bail out. It would have been down over the jetstream, though, and I knew I didn't have a chance. So I made my flare fully expecting the airplane to snap and do everything it's not supposed to do. Lo and behold, it flared out like a Piper Cub."

We found out what was wrong. The cannon had exploded, and the wing flaps had failed because they lost electrical power. The dead-stick landing with an F-104 is quite an adventure, because as heavy as it is, with that small wing and no power, the pilot must be precise in putting the airplane down. LeVier had studied the figures for such a landing in advance, knew exactly what he should do, did just that, and made a perfect dead-stick landing.

In his case, Herman had no choice but to leave his airplane. He was wearing a pressure suit for gun-firing tests at supersonic speed above 50,000 feet when he felt a rush of air and his faceplate froze over. He couldn't see any instruments but knew he did not want to disconnect the pressure suit from the aircraft at that high altitude. So he started counting to himself, waiting as long as he felt he could before raising the faceplate.

"The only thing I was really interested in was the altimeter," he explained. "I didn't care where anything else was."

Herman wanted to help us get all the anwers we could to the cause of that accident so he volunteered to take sodium pentothal administered by Lockheed medical director Dr. Charles I. Barron. From his descriptions we found out exactly what happened.

An explosion from accumulated gun gas had blown open one of the landing wheel well doors, admitting a rush of upper atmospheric cold. Having been an exhibition parachute jumper in his early career allowed "Fish" to take this bailout in stride. And the gun-firing problem soon was solved.

The F-104 had been designed as an interceptor-fighter, an assignment for which the plane was very, very good. But when the NATO countries decided they wanted to build and use them, they wanted additional performance—low-level ground attack as well. We were able to double the airplane weight from the original 16,500 to 33,000 pounds without any change in wing area—which is less than 200 square feet. To do this, takeoff speed had to go way up. The airplane became a hotrod of the first order. And it was being flown in some of the world's worst weather and terrain. Some countries—Norway, Canada, and Taiwan, for example—set safety records for fighter aircraft with it.

For special reasons, West Germany had problems with the airplane. The Germans had a very sophisticated version—very high performance combining interceptor, bomber, and reconnaissance capability in the aircraft. Later models added infrared gunsight and inertial navigation.

The trouble was that for ten years after World War II, German pilots had not had modern jet experience, especially supersonic. Nor were conditions conducive to keeping trained pilots and mechanics in the air force. While there was a great deal of press attention to the initial high accident rate with the F-104s, it went virtually unnoticed that the West Germans earlier had bought a number of F-84s and lost about 40 percent of that fleet in a very short time.

It wasn't the equipment but the nature of the operation that was the basic problem. Finally, West Germany arranged to train their pilots at Luke Air Force Base in Arizona, where weather permitted year-round flying. Properly trained in an air force that offered a stable career, the West German pilots achieved an excellent in-service record with the airplane.

In its first year of service with the U.S. Air Force, the F-104 set some significant official records. In 1958, it recaptured the world altitude record for this country at 91,243 feet. That same year it set a new speed mark of 1,404.19 miles an hour. It established seven records for time-to-climb to altitude. In 1959, it set a new altitude record of 103,395.5 feet. The F-104 won for the Air Force, GE, and Lockheed the Collier Trophy in 1959, for the previous year's "greatest achievement in aviation."

The F-104 later became also an aerospace trainer for the USAF, simulating re-entry and zero gravity conditions for astronauts at the Aerospace Research Pilot School at Edwards AFB.

When the MiG-21—a high-performance Russian jet—began to appear in increasing numbers with Soviet bloc air forces, replacing the earlier MiG-17 and -19, Western Europe began to look toward a new air-superiority fighter. Their principal air defense weapons were Starfighters, Phantoms, and Lightnings. Intelligence reports and news items indicated that advanced fighter aircraft in East European countries outnumbered NATO air forces by as much as six to one.

As a follow-on airplane to the F-104 for our European allies, Lockheed developed a very practical and productive proposal. Use the expensive, proven components and systems from the F-104 but add a larger wing and new tail and increase the power for an all-around fighter. The new design promised to outmaneuver all other aircraft known to be flying including the MiG-21. We called the airplane the "Lancer."

It was proposed with a choice of engines, the familiar GE model or a new advanced-technology design from Pratt & Whitney with a Mach 2.5 speed. We proposed to conduct development and initial flight tests at the Skunk Works in cooperation with the European engineers, and that production of the plane be programmed entirely in Europe. It could have meant millions of dollars saved in production costs as well as jobs for many thousands of people in the plants where they had turned out the F-104.

The international competition among manufacturers for sale of a new airplane to the NATO countries was intense. France's Dassault proposed an advanced version of its Mirage F-1.

Two other U.S. manufacturers were after the business. McDonnell-Douglas offered a modification of its twin-engine Phantom, to be designated F-4F. Northrop proposed a totally new design, P-530 (F-5), which would not have been available until 1976. Lockheed promised the Lancer for service in 1973. By that year, no decision had yet been reached, and we still were trying to sell the Lancer overseas. The airplane was not proposed to the USAF, which was interested in developing a new aircraft.

As a sidelight on international sales of aircraft, on the very day that we started touring Europe with our proposals for the Lancer, one of our competitors set out on the same circuit to sell theirs. Months later we discovered that both of us had retained the same overseas marketing consultant. He was being paid by both sides.

The Air Force in this country was considering development of two new fighters, the F-14 and F-15.

Back in 1969, I questioned publicly whether these aircraft actually would be competitive with the best Russian fighters. Specifically, I said that I thought the cost of the proposed F-15 would be more expensive than necessary, that a smaller, less-expensive airplane could do the job, just as well.

Stuart Symington, former Secretary of the Air Force, then Senator from Missouri, very much wanted that F-15 contract for his state. He called in Dan Haughton, then Lockheed board chairman, and me and announced that whether we liked it or not that contract for the F-15 would be awarded to McDonnell. Kelly Johnson was not to give any more argument. Haughton was under the gun and promised that he'd see that I didn't. I did not promise in my own right.

We, Lockheed, had made an unsolicited proposal to the USAF for an advanced, highly-maneuverable lightweight fighter that we could have had flying within a year at absolute minimum cost. We had lined up a dozen of our vendors with whom we were then working on another project, including Pratt & Whitney for the jet engines, other suppliers for armament, gunsights, wheels, tires—the whole package. It was a darn good airplane. The X-27, later designated X-29, was basically a new airplane, but it utilized the nose design of the F-104 which by that time had fired millions of rounds of ammunition. We even proposed firing tests on the first flight to prove we had a fighting airplane.

David Packard, then Deputy Defense Secretary, was very much impressed with the proposal. But Robert Seamans, Jr., Air Force Secretary, did not like the idea of buying a fighter in that manner. He preferred the conventional method—an experimental model first, production plans later.

I disagreed with the USAF on procurement policy.

"This airplane is not so advanced that you cannot develop the 'X,' the experimental airplane, into a production prototype," I argued. "I don't want to draw a line on paper that does not consider production. Why go to double prototyping?"

We came close to receiving a contract for that airplane, but what eventually killed any prospect for our producing the lightweight fighter were the financial problems that Lockheed encountered in 1971—first, losses from several fixed-price contracts for the U.S. military, then the threatened loss of the company's new L-1011 commercial transport program with the unexpected bankruptcy of Great Britain's Rolls-Royce, manufacturer of the engine. The very future of Lockheed was in question, and the Air Force reasoned understandably that they should not risk awarding the contract to the company.

Our proposal did, I believe, result in the USAF's eventual design competition for a lightweight fighter. The plane they got at least ten years later, after double prototyping by General Dynamics and at nearly three times the cost, was comparable in performance. That was the F-16.

If the military would spend one or two percent of the cost of developing an experimental airplane in planning production at the same time, it would come back in savings many, many times over.

The Skunk Works method of developing the F-104, for example, could save a considerable amount of money if applied to procurement of new aircraft.

On that airplane, every time we released an engineering drawing to our manufacturing director, Art Viereck, to build a part for the experimental airplane, we also released it to a group of production engineers with these instructions: "Find every alternative way of making this, ruling out adverse effect on drag, maintenance, or cost. You can affect them all favorably."

When we finished building the prototype, we had a thick report on how to build a production model. We sat down for three days with that book to choose the best way to build the airplane from every point of view. It saved from $10,000 to

Kelly with the first JetStar, which critics declared would never attain commercial success.

$20,000 for every airplane built, by my estimate. Total production worldwide was about 2,500.

Costs must be considered. Aircraft are getting to be so expensive they hardly are worth it for what they can do. With the price of fighter aircraft now running more than $30 million per plane with all the equipment, not including pilot costs, I can foresee the day when the fighter pilot will be on the ground, flying an unmanned fighter with a missile in it. With the latest electronic advances, I think this can be done remotely at a great saving in aircraft costs—and, of course, great saving in manpower, to say nothing of the greater safety for the pilot. It's worth considering.

13

Working with "Spooks"

WHATEVER ELSE IT HAS DONE AND EVER WILL DO, the U-2 is indelibly identified in the public mind as the "Spy Plane" in which Francis Gary Powers was shot down over Russia on May Day of 1960 while on a photo reconnaissance mission for the CIA.

The plane has been used for high-altitude weather research, earth resources survey, communications satellite, and aerial mapping—as well as reconnaissance.

It came into being for that purpose—reconnaissance—though this was disguised in first public announcements of its existence.

In a press release for Monday, May 7, 1956, the National Advisory Committee for Aeronautics (NACA) announced "Start of a new research program" and a "new airplane, the Lockheed U-2 . . . expected to reach 10-mile-high altitudes as a matter of routine."

"Tomorrow's jet transports will be flying air routes girdling the earth . . . at altitudes far higher than presently used except by a few military aircraft," NACA Director Dr. Hugh L. Dryden explained. "The availability of a new type of airplane . . . helps to obtain the needed data . . . about gust-meteorological conditions to be found at high altitude . . . in an economical and expeditious manner."

The NACA release identified the plane with specific high-altitude research work in clear air turbulence, convective clouds, wind sheer, and jet stream. Also to be studied were cosmic rays and concentration of certain elements in the atmosphere including ozone and water vapor.

"As a result of information so to be gained, tomorrow's air travelers might expect a degree of speed, safety and comfort beyond present hope of the air transport operators," the announcement continued.

"A few of the Lockheed airplanes are being made available for the expanded NACA program by the USAF.

"The first data, covering conditions in the Rocky Mountain area, are being obtained from flights from Watertown Strip, Nevada."

All true—in time.

An internal Lockheed memo issued at the same time to corporate executives from Courtlandt Gross described the plane as "of conventional, straight-wing design . . . with light wing loading to enable routine flight in the 50,000– to 55,000- foot altitude range . . . powered with a single Pratt & Whitney J-57 engine.

"We built a prototype with our own funds, and its high-altitude capabilities quickly attracted sufficient military interest to earn us an Air Force contract for a limited number. . . .

"Its development has been . . . under the direction of C. L. Johnson, who last month assumed the new positions of vice president research and development, and director of special projects for the California division. . . .

"The Air Force found the U-2 to be a good, economical flight platform for use in a joint test program with the Atomic Energy Commission. For this reason, and because of the experimental nature of the aircraft and its test equipment, further details are classified."

On July 9, another NACA press release covered the U-2's assignment overseas.

"High Altitude Research Program Proves Valuable," was the headline.

"Initial data about gust-meterological conditions to be found at 10-mile-high altitudes which have been obtained to date by the relatively few flights of Lockheed U-2 airplanes have already proved the value of the aircraft for this purpose. . . .

". . . Within recent weeks, preliminary data-gathering flights have been made from an Air Force base at Lakenheath, England, where the Air Weather Service of the USAF is providing logistical and technical support. As the program continues, flights will be made in other parts of the world."

The six-page news release went on to list and describe the atmospheric data-gathering instrumentation carried in the airplane.

Design of the U-2 had begun several years earlier. In 1953, we at Lockheed had been made aware of this country's desperate need for reconnaissance of Soviet missile and other military capabilities. A requirement existed for an airplane that could safely overfly the USSR and return with useful data. The plane was needed "now."

My first thought was to explore the proven F-104 design for possible application to this mission. Phil Colman and Gene Frost of our preliminary design department were assigned this task. It soon became obvious that the only equipment we might retain from the F-104 might be the rudder pedals. We initiated an entirely new design.

These were the requirements. The airplane would have to fly at an altitude above 70,000 feet so vapor trails would not give away its presence, have a range better than 4,000 miles, have exceptionally fine flight characteristics, and provide a steady platform for photography with great accuracy from this high altitude.

It would have to be able to carry the best and latest cameras as well as all kinds of electronic gear for its own navigation, communication, and safety.

Our first presentation was to the Air Force, where it was turned down as too optimistic. They questioned that any engine even would operate at the altitude we were proposing. They were correct in that there was not proof at that time that this was possible. And the Air Force already had an airplane in development with the Martin company—an airplane with two engines which was preferred to our single-engine design.

But our proposal reached Trevor Gardner, then Assistant Secretary of the Air Force for Research and Development, and a brilliant engineer in his own right. Late in 1953, he invited me to come to Washington to discuss it. He had assembled a committee of scientists and engineers, and for three days they put me through a grilling as I had not had since college exams. They covered every phase of the aircraft design and performance—stability, control, power plants, fuels—everything.

Later I met and lunched with Air Force Secretary Harold Talbott, CIA Director Allen Dulles, and his right-hand man, Larry Houston, among a distinguished group. When I was asked why I thought Lockheed could do what I proposed—build 20 airplanes with spares for roughly $22 million and have the first one flying within eight months, Gen. Donald Putt graciously volunteered, "He has proven it three times already—on the F-80, F-80A, and F-104."

The secrecy of this project was impressed on me by Gardner. I understood that I was essentially being drafted for the job—becoming a "spook"—the intelligence community's label for their agents. I returned to Burbank with instructions to talk only with Robert Gross and Hall Hibbard. Despite the fact that they had sent me to Washington with instructions not to commit to any new projects because the plant already had several military programs in engineering, they agreed that we must cooperate with this important work. Trevor Gardner himself later met with Gross and Hibbard to confirm formally the contract.

I organized the project with 25 engineers including myself in the experimental department and with Art Viereck again in charge of the shop, with a staff that gradually grew to 81 people.

It had been decided in Washington that the project could best be handled under the CIA's direction and funding, with the Air Force providing the engine. We had been working with Pratt & Whitney on developing the J-57 for this purpose. There was no time to develop a new engine; we had to use existing equipment.

Richard Bissell, "special assistant" to Dulles, was selected to direct the program. Bissell, an economist, quickly became very knowledgeable on engineering matters. He has described his introduction to the program:

"I was summoned one afternoon into Allen's office; and I was told with absolutely no prior warning or knowledge that one day previously President Eisenhower approved a project involving the development of an extremely high-altitude aircraft to be used for surveillance and intelligence collection over 'denied areas' in Europe, Russia, and elsewhere. I was to go over to the Pentagon, present myself in Trevor Gardner's office and there with Gardner, Gen. Donald Putt of the Air Force, Gen. Clarence Irvine and others, we were to decide how the project was to be organized and run.

"The first time I heard Kelly's name mentioned was in a call put through by Trevor Gardner to Kelly in which he gave him a go-ahead to develop and produce 20 U-2 aircraft. We had an almost impossible schedule to meet."

First program manager for the USAF was Col. O. J. "Ozzie" Ritland, special assistant to Deputy Chief of Staff, Development, USAF Headquarters. He was followed by Col. Leo P. Geary, USAF.

One of our first tasks was to find a base from which to operate. The Air Force and CIA did not want the airplane flown from Edwards AFB or our Palmdale plant in the Mojave Desert. So we surveyed a lot of territory. There are many dry lakes in and around Nevada, and the lakebeds are generally quite hard, even under water in the rainy season. A site near the nuclear proving grounds seemed ideal, and Bissell was able to secure a presidential action adding the area to the Atomic Energy Commission's territory to insure complete security.

Dorsey Kammerer and I flew to what would be the test base. I had an Air Force compass, and he had some surveying equipment for use on the ground. Kicking away some of the empty. 50-caliber shell cases and other remnants of target practice, we laid out the direction of our first runway.

A road had to be laid out, hangars constructed, office and living accommodations built, and other facilities provided.

Since Lockheed did not have a license to build on the nuclear proving ground, we gave our drawings to a contractor who did. When he put the work out for bid, he was told by one company, "You want to look out for this 'CLJ' outfit. We've looked them up in Dun & Bradstreet and they don't even have a credit rating." We were using my initials for identification.

By July, "agency" personnel, Air Force, and Lockheed people began to move in.

In recruiting mechanics and technical people to work on the project, we named it The Angel from Paradise Ranch—Angel because it was such a high-flying airplane, of course, and Paradise Ranch because we thought that would attract people. It was kind of a dirty trick since Paradise Ranch was a dry lake where quarter-inch rocks blew around every afternoon.

Actually, in the Skunk Works, we've never had trouble getting workmen to go wherever we've needed them because they know wherever it is the work will be exciting and challenging. And where it's really rough, they are paid a bonus of perhaps 15 percent on top of a good basic salary, plus living expenses. They cannot, of course, take their families with them for security reasons, but they can return home at least once a year when on long assignments.

Security was so strict that after we had submitted our first vouchers for progress payment on the contract, two checks for a total of $1,256,000 arrived in the mailbox at my Encino home. It seemed prudent to establish a special bank account after that.

One of the Skunk Works rules that we actually make contractual is that funding should be timely. We plot our way along the program, report progress monthly and the amount of costs. We require incremental payment so we don't have to go running to the bank to carry the government. But on a couple of occasions we have.

On the U-2 program, Colonel Geary once came to tell me the money had not yet been appropriated for the next 30 days. Well, I had read in the newspaper about delayed budget approvals, and had already made arrangements for us to borrow the money—as I recall it was two or three million dollars. But interest rates then were only about five percent.

The government got a bargain on that contract when completed—about $2 million in refunds on contract costs, and six extra airplanes from spare parts we didn't need because the U-2 functioned so well.

Our security on the U-2 was threatened at one point in a horrendous example of bureaucratic mixup—understandable because of that very secrecy. The Air Force in 1955 issued a proposal to industry for a weapon system designated the X-17. It appeared that they copied our proposal; it was a dead ringer for our original presentation. I phoned Bissell on a Sunday night after I'd discovered this to point out the breach of security involved. When I showed the specifications to him and Ritland the next week in Washington, they reacted with stark horror. We went to Gardner's office; he went to Talbott. The proposal was withdrawn within a few minutes. It would have had the Air Force spending from three to six million dollars when actually a better airplane—the U-2—would have been flying before they had design proposals returned from the aircraft companies.

By July of that year, "the ranch" was ready. The first U-2 was disassembled and at 4:30 a.m. on July 24, we arrived at the plant to begin loading the plane into a C-124 for the flight to Nevada. We would follow in a C-47.

The base commander, however, denied permission for the C-124 to land because the normal tire pressures would be too high for the thin surface of the runway. But we determined it would be safe if we let most of the air out of the tires, providing about four times the tire footprint area on landing. It was a novel solution, not covered in operating procedure for the base, and the commander couldn't and wouldn't approve it. So we phoned Washington for permission to overrule the local authority, reduced the air pressure, took off for Nevada, and made a beautiful landing on the soft runway. I measured the impact area myself, and the deepest impression was one-eighth of an inch. Had we not devised this unusual procedure, we would have had to land elsewhere, haul the plane over roads not in very good shape, and miss the target flight date by as much as a week.

Our first flight was unprogrammed. It was supposed to be a taxi test with Tony LeVier at the controls. The airplane was so light that on his second taxi run it just lifted to a height of about 35 feet before Tony realized he was off the ground. And when he tried to land, the darned airplane didn't want to. It could fly at idle power on the engine. He managed to bounce it down, and in the process bent the tail gear a bit. But we soon had it fixed. The unofficial first flight was on August 4, 1955.

Tony made his first flight to 8,000 feet in a rainstorm. The lakebed was dry when he took off, but he flew through rain just north of the field. I was observing from the C-47 "chase" plane. The airplane flew beautifully, but again Tony had trouble on landing. He came in tail high and the plane porpoised badly. He made five other attempts before I could talk him down. We discovered the airplane makes a fine landing when the tail wheel hits at the same time or slightly ahead of the main gear. The landing characteristics are quite unusual but not unexpected.

Ten minutes after the U-2 landing, the lake was flooded with two inches of water. This was incredible, because average annual rainfall for the previous five years had been 4.3 inches. We had had almost that much already within the preceding two weeks.

That night all of us celebrated with the usual beer and arm-wrestling contests. Thanks to my early lathing work, I was pretty good at both. It has been our policy in the Skunk Works that everyone—all the workers as well as engineers and executives—sees the first flight and is included in the traditional party afterward.

To witness the "official" first flight on August 8, we invited our customers from Washington. It was a very successful flight to 35,000 feet.

From then on it was drive, drive, drive. Build the airplanes, get them in operation, train ground and flight crews, maintenance men, military pilots.

With the developing U-2, an aircraft considered a technological triumph by any standard. Below, the high-flyer in Air Force markings.

It wasn't long—eight or nine months after first flight—that the U-2s became operational. Two of them were deployed to Lakenheath in beautiful country northeast of London. They were spotted by the British Air Defense Force while on practice flights at 70,000 feet, and the RAF scrambled to try to reach them to investigate.

The British and Russians at that time were at odds over an incident in which the Russians had killed a British frogman inspecting a Russian cruiser in harbor in England. The presence of the U-2s could only worsen an already sensitive situation. They were removed to West Germany in June of 1956, for their first flights directly over Moscow and Leningrad. "A brilliantly successful first field operation," as described later by Bissell.

There were diplomatic protests after a few overflights, and missions were curtailed a while and then resumed. The points of origin varied, and the U-2 flights over Russia, or near enough to its borders for reconnaissance, continued for about four years until Powers was shot down.

He had taken off from Pakistan, and was to fly over Russia to Norway. He was flying a U-2C, with a new improved Pratt & Whitney engine that gave it 3,000 to 5,000 feet more altitude than the original U-2A. Photos from earlier overflights had shown as many as 35 Russian fighters trying to climb up to intercept the U-2. They formed an aluminum cloud over the terrain to be photographed.

During the period of U-2 overflights, the Russians had been working diligently to improve their SA-2 missile and radar systems. When Soviet leader Nikita Khrushchev triumphantly announced that Powers had been shot down west of Sverdlovsk, we tried to reconstruct what had happened. We simulated Powers' flight mission and studied what aircraft components might fail and cause him to lose cruising altitude. We found nothing in the aircraft or its systems likely to lead us to doubt that the aircraft had been hit at altitude as the Russians stated.

But they had published a photo of the downed airplane. I

knew it was not a U-2 and believed it was one of their own fighters they had shot down while trying to get the U-2. What they exhibited obviously had been scraped up with a caterpillar blade and had school kids running all over it. It didn't take any stroke of genius to know that they wouldn't have handled a captured U-2 like that, allowing kids to play on it. The CIA wanted me to provoke, insult actually, the Russians into revealing more about the incident. So, I challenged them publicly.

"Hell, no," I was quoted in the press, "That's no U-2." At about that time it was suggested to me by security people that I not go to work and come home always by the same route or establish regular patterns of movement. And for a few years during this and other secret aircraft developments, I slept with an automatic pistol close by.

The provocative strategy worked. The Russians put the real U-2 on display in Moscow. From the many excellent news correspondents' photos, particularly the complete and high quality coverage by *Life* magazine's Carl Mydans, we could determine many things. Among them:

Both wings failed because of down-bending, not penetration of critical structure by shrapnel from a missile.

None of the pictures showed a horizontal tail. And the right section of the stabilizer was missing. While this damage is conceivable from a crash landing, it was improbable because of the relatively undamaged condition of the vertical tail itself.

The design of the U-2 wing is so very highly cambered that without a tail surface to balance the very high pitching moment, the aircraft immediately goes over on its back; and in severe cases the wings have broken off in down-bending. This occurred once in early testing when the pilot inadvertently extended wing flaps at high cruise speed, resulting in horizontal tail failure. This takes place in a few seconds, at great acceleration and with the fuselage generally spinning inverted.

When Powers was exchanged in February 1962 for a Russian spy, I met and talked with him as soon as possible. His statements matched our conclusions.

Between what we had deduced and what Gary told us, it

appeared that an SA-2 missile had knocked off the right-hand stabilizer while he was at cruising altitude. The airplane then, predictably, immediately went over on its back at high speed and the wings broke off in downbending. Gary was left sitting in the fuselage with a part of the tail and nothing else. He did not use the ejection seat, but opened the canopy to get out.

With the airplane spinning badly and hanging onto the windshield for support, he tried to reach the destruct button to destroy the airplane. It was timed to go off about ten seconds after pilot ejection. But he could not reach the switch. We simulated the situation and it just was not possible with the forces on his body. He had to let go. His biggest worry then was that the fuselage and flailing tail would descend through his chute. But he landed uninjured in a farming area and was captured almost immediately.

The following is the conclusion of my report to our government on February 21, 1962.

"I was so impressed by the very clear description of the incident by Frank, and having direct knowledge of what he was ordered to do in case of capture, that I will gladly contribute to a fund for decorating this officer for the fine job he did under the most difficult circumstances. He satisfied me, by detailed questions, that the Russians could not have brainwashed him on detail matters of his escape from the aircraft."

Powers came to work for us at the Skunk Works.

After the U-2 was taken off Soviet overflights, it was assigned many other jobs. But it had developed very important information and it would be many years before we had anything to surpass the U-2's photographic capability—and then it was our own Lockheed Agena satellite and the SR-71 aircraft.

The airplane is a good workhorse, really earning its "U" designation for utility. It even has done the high-altitude weather research first announced for it—in fact, a great deal of atmospheric and earth resources surveying for NASA. For that assignment, the ER-2, for "earth resources," is their latest model, flying higher and farther than its predecessor U-2.

The U-2C has operated on the carrier *Kittyhawk* and others

SR-71 production at Lockheed's "Skunk Works" in Burbank, Calif.

in demonstrations for the Navy. A later model, the U-2R—for "revised"—would be even better for carrier operation, although its 100-foot wingspan versus the earlier 80 feet does take up a lot of space.

By about 1965, the U-2s had served nearly ten years and were showing their age. We proposed to build the U-2R. The Pratt & Whitney J-57 engine had achieved a 20 percent increase in power, so we could fly with a heavier load. We had to reduce the wing loading while keeping the span as great as we could; we extended the wing to that 100-foot dimension, providing 1,000 square feet of wing area instead of 600 sq. ft. The cockpit was enlarged 45 percent, which made it more comfortable on long missions. There were other improvements to equipment and field serviceability.

The airplane operated in Southeast Asia over a period of four years with a maintenance readiness record of dispatch within 15 minutes of assigned time on 98 percent of the calls.

The availability was as good as for the big widebody airliners. And this was in snake-infested jungle areas with the absolute minimum number of maintenance personnel and equipment.

When the *Mayaguez* was captured and our communications satellite went out of commission, one U-2R stayed at altitude for a total of more than 27 hours in a three-day period acting as a communications link for our military forces.

Our B-52 pilots in the war theater preferred not to go out until the U-2 had told them who was coming at them.

The "R" can fly almost nine miles a minute above 70,000 feet and search 300 miles to the horizon. It can pick up enemy radar and communications. The plane can stay out for as long as ten hours, making it a very good early warning aircraft. Used with the Navy, it could land on a carrier to refuel when ranging far from land bases and be off again. One day, I think the Navy may want to use the U-2 for that purpose.

The latest U-2 version in Air Force service is the TR-1, for "tactical reconnaissance." It was named on the spot by USAF Gen. David C. Jones, then Chairman, Joint Chiefs of Staff. "We have to get the U-2 name off the plane." Even the military is sensitive to that "Spy Plane" image. "We'll call it the TR-1, tactical reconnaissance one."

The TR-1 entered service with the Air Force in 1981. Both the TR-1 and ER-2, latest U-2 versions, are 40 percent larger than the original. Wing span is 103 feet; fuselage, 63 feet long. The contract for both was a cooperative effort by the USAF and NASA. The ER-2 is equipped with interchangeable noses for varying missions. Without penetrating foreign territory, the TR-1 with highly sophisticated sensors operating from very high altitude can identify targets and threats behind enemy lines.

Development of the latest long-endurance capability, operating now in the TR-1, occupied us for many years. There are many more things we can do now that we could not do when the plane first was developed—nearly 30 years ago! There are many new technologies—silicon chips, optical fibers, various composite materials. But I don't think anything is going to fly higher subsonically. And that, of course, is why we went to our next project, the SR-71, to get more altitude and speed. It was under design as soon as the U-2 became operational.

"About a year after that first flight in 1956," Bissell has said. "I came to the conclusion that we should start working on the successor to the U-2, because it was clear to me that sooner or later the U-2 would be vulnerable to interception."

14

Blackbirds Fly Stealthily— Three Times the Speed of Sound

$F_{EBRUARY}$ 29, 1964. PRESIDENT REVEALS SECRET AIR-CRAFT:

"The United States has successfully developed an advanced experimental jet aircraft, the A-11, which has been tested in sustained flight at more than 2,000 mph and at altitudes in excess of 70,000 feet," President Lyndon Johnson announced today. . . .

"The performance of the A-11 far exceeds that of any other aircraft in the world today. . . .

"The project was first started in 1958. . . . The Lockheed Aircraft Corporation at Burbank, California is the manufacturer of the aircraft. The aircraft engine, the J-58, was designed and built by the Pratt & Whitney Division, United Aircraft Corporation. The experimental fire control and air-to-air missile system for the A-11 was developed by the Hughes Aircraft Company.

"In view of the continuing importance of these developments to our national security, the detailed performance of the A-11 will remain strictly classified and all individuals associated with the program have been directed to refrain from making any further disclosure. . . ."

JULY 24, 1964. PRESIDENT ANNOUNCES NEW SPY PLANE

President Lyndon Johnson at a press conference: "I would

like to announce the successful development of a major new strategic manned aircraft system which will be employed by the Strategic Air Command. This system employs the new SR-71 aircraft and provides a long-range advanced strategic reconnaissance plane for military use capable of world-wide reconnaissance for military operations. . . .

"The SR-71 aircraft reconnaissance system is the most advanced in the world. The aircraft will fly at more than three times the speed of sound. It will operate at altitudes in excess of 80,000 feet. It will use the most advanced observation equipment of all times in the world . . . an outstanding reconnaissance capability. . . . The SR-71 uses the same J-58 engine as the experimental interceptor previously announced but it is substantially heavier and it has a longer range. The considerably heavier gross weight permits it to accommodate the multiple reconnaissance sensors needed by the Strategic Air Command to accomplish the strategic reconnaissance mission in a military environment.

"This billion dollar program was initiated in February 1963. The first operational aircraft will begin flight testing in early 1965, and deployment of production units to the Strategic Air Command will begin shortly thereafter. . . ."

DECEMBER 23, 1964. DEPARTMENT OF DEFENSE NEWS ANNOUNCEMENT

"First flight of the U.S. Air Force SR-71, the new long-range strategic reconnaissance aircraft, took place yesterday at Palmdale, California.

"The Lockheed-built aircraft, flown by company test pilot Robert Gilliland, was in the air approximately one hour.

". . . All test objectives were met.

"The aircraft will be assigned to the Strategic Air Command at Beale Air Force Base, Marysville, California, in 1965."

Almost as soon as the U-2 was in operation, we began to plan its successor.

Efforts to improve the U-2 had yielded very little gain. A new design would be necessary. We set up our own requirements. There were none officially yet—just the need. We knew we needed more altitude and, especially, more speed.

Vulnerability studies led us to the decision that the next airplane should operate at altitudes well over 80,000 to 85,000 feet, fly at speeds well over Mach 3, and be able to out-maneuver any SA-2 missile the Russians might develop. It had to be stable enough in flight to take a good photograph from altitudes above 90,000 feet. It had to retain the characteristics of the U-2—be able to photograph very, very small targets on the ground—while flying four to five times as fast. We wanted it to have global range—with multiple midair refuelings from the KC-135 aerial tankers. The aircraft also should present an extremely low radar cross-section—be very difficult to detect.

The decision to assign this project to Lockheed was arrived at only after consideration of several design proposals from other companies. From April 21, 1958, through September 1, 1959, I made a series of proposals for Mach 3-plus reconnaissance aircraft to Richard Bissell of the CIA and to the USAF. Bissell, as chairman of the review committee, was reluctant to award the contract for its successor to the same company that had built the U-2 without a design competition.

Some of the other entries were interesting.

A Navy in-house concept proposed an inflatable rubber vehicle which could be carried to altitude by a balloon, then boosted by rocket power to a speed where its own ramjets could take over. This rapidly was demonstrated to be unfeasible. The balloon would have had to be a mile in diameter to lift the unit, which itself had a proposed wing area of one-seventh of an acre.

Convair proposed a ramjet-powered Mach 4 aircraft, which also had to be carried aloft by another vehicle and launched at supersonic speeds where the ramjet power could take over. Unfortunately, the launch vehicle was the B-58, which could not attain supersonic speed with the bird in place. Even if it could, the survivability of the piloted vehicle was in question because of probable ramjet blowout in maneuvers.

The total flight time for the Marquardt ramjet at the time was not over seven hours, obtained mainly on the ramjet test vehicle for the Boeing Bomarc missile. This test vehicle, the X-7, had been built and operated by the Skunk Works.

On August 29, 1959, our A-12 design, the twelfth in the series, was declared the winner and Bissell gave us a limited go-ahead. We were to conduct tests on models, build a full-scale mockup, and investigate specific electronic features of the airplane over a four-month period.

On January 30, 1960, we were given a full go-ahead for design, manufacture, and test of 12 aircraft.

The code name was Oxcart, a name selected from a list of deliberately deceptive identifications. Obviously an oxcart is a slow-moving body. This CIA program led to other versions of the design for the U.S. Air Force. The next would be a long-range fighter for the Air Defense Command, first discussed with Gen. Hal Estes on March 16 and 17 of that same year. It became the YF-12A.

In January 1961, I made a proposal for a strategic reconnaissance airplane to Air Force Secretary Dr. Joseph Charyk, U-2 project officer Col. Leo Geary, and Lew Meyer, an Air Force finance officer. This encountered initial opposition in some quarters where it was seen as competition for funds for the North American B-70 program, then the object of considerable controversy. But it became the SR-71, the prime reconnaissance airplane for the Air Force today. There also would be a fourth verson—the D-21, an unmanned drone.

The aircraft that were to become the "Blackbirds" were the first to use the "Stealth" technology we developed for radar avoidance. We had tried to work it into the U-2 as a modification after the aircraft already were in operation and got nowhere. To be "stealthy," aircraft must have certain features incorporated in design from the very beginning, not added later. "Stealth" must be designed into the airplane from the first three-view drawing to be effective. This is what we did with the Mach 3-plus aircraft.

We had achieved Mach 4 flight earlier for a few seconds with the X-17, our ramjet test vehicle—before that the X-7. The idea of attaining and staying at Mach 3.2 over long flights was the toughest job the Skunk Works ever had and the most difficult of my career. Early in the development stage, I promised $50 to anyone who could find anything easy to do. I might as well have offered $1,000 because I still have the money.

Aircraft operating at those speeds and altitudes would require development of special fuels, structural materials, manufacturing tools and techniques, hydraulic fluid, fuel-tank sealants, paints, plastics, wiring and connecting plugs, as well as basic aircraft and engine design. Everything about the aircraft had to be invented. Everything.

In the search for a suitable fuel we considered initially such exotic candidates as liquid hydrogen, coal slurries, and boron slurries.

Design of a liquid hydrogen powered airplane actually was carried to the point approaching a production contract for a substantial number of them. The concept had seemed promising. While the airplane was very large, it was very light in weight. Powered by special engines developed by Pratt & Whitney, it would have a cruising altitude well above 100,000 feet—higher than the Blackbirds later achieved, although with less speed and range. But the further we went into development, the worse the problems we perceived.

The CL-400 was essentially a big flying vacuum bottle, with the liquid hydrogen heavily insulated to keep it at very low temperatures and as close to absolute zero as possible. The problems of transporting this fuel from a plant in the United States to wherever the airplane could be based would have been prohibitive. It would have required a whole fleet of C-124s to keep just a few liquid hydrogen planes flying. No foreign nation was likely to approve our flying in with such quantities of it nor to allow us to put a ship in their harbors and liquefy it on the site.

We had engaged the Pomeroy Company to design a hydrogen liquefaction plant that was to have been erected near our aircraft plants and the Palmdale Airport in the Mojave Desert. It would have used ten percent of the natural gas input to the city of Los Angeles in 1972 and '73!

One day we were visited at Lockheed by Assistant Air Force Secretary James Douglas and Gen. Clarence Irvine of the Air Force. Their question was, "How much stretch have you got in this thing, Kelly?"

"Let's take a look at it," I said. "Here's the inboard view. You can see it's totally liquid hydrogen from one end to the other except for a small cockpit up front."

You do not put liquid hydrogen in nooks and crannies and odd-shaped tanks. The container has to be cylindrical—and very well insulated. With this airplane, we didn't have the condition we've always had with other aircraft, both piston-powered and jets, where extra fuel could be added for a little more power and range. The Constellation gross weight, for example, doubled in its lifetime. We were able to do the same with most of the fighters, too. But with the liquid hydrogen airplane, once you set down the tank volume, that's it. You could carry external tanks, but it would be difficult, and the airplane would carry added drag because of air resistance.

So the Secretary and the General turned to Perry Pratt, head of Pratt & Whitney engine design. "Maybe there's something in the engine. Perry, how much stretch do you have in this engine?"

"Perhaps three to four percent in five years," was the answer.

Overall, it wasn't a very good forecast. We all agreed to cancel that effort without any more expenditure of funds. And had we proceeded, we would have run right into the energy crunch of the 1970s.

Coal slurries—finely-ground-up coal mixed with a light oil base and water—were a possible power source, injected into the engines as fuel. But the tiny coal cinders tended to ruin the turbine blades.

Boron compounds—in slurries—were tried, too. But these were difficult to use and plugged up the injector nozzles, not only in the engine but afterburners as well.

We decided to stick with liquid petroleum as fuel. It would have to be a very special fuel, though, for those operating altitudes and temperatures—from −90°F for midair refueling at high altitude to +650°F in supersonic flight.

We put the problem to our old friend, Jimmy Doolittle, now a top executive with Shell Oil. That company had come up with what we called LF-1A, "Lockheed Lighter Fluid 1 for the U-2; and it was a good fuel for that airplane. Shell came through again, with cooperation from the Ashland and Monsanto companies and Pratt & Whitney, with a new chemical lubricant and fuel for the Mach 3-plus aircraft. This we call LF-2A. The Air Force has its own designations, of course.

Development of that fuel took a lot of doing. It was very expensive, but it's an excellent fuel.

The fuel in this design also acts as an insulating element. The fuel tanks not only contain the fuel but are constructed to protect the landing gear. The gear retract up into the middle of the tanks. With radiant cooling to the fuel, the rubber tires are insulated against the very high temperatures the plane encounters during hour after hour of flight. And the plane doesn't land with the tires ready to blow out.

In selection of structural materials for a Mach 3-plus airplane, aluminum automatically was ruled out. It would not withstand the ram-air temperatures of 800°F over the body of the plane. High-strength stainless steel alloys or titanium would be required for the basic structure. And high-temperature plastics would have to be developed for radomes, cockpits, and certain other areas.

Stainless steel actually is a better high-temperature material than titanium. But visiting the Lockheed-Georgia plant, which was building parts for the supersonic B-70 bomber, I saw what it took to make basic honeycomb panels for the fuselage: a "clean-room" environment—what was essentially a big pressurized airbag—with pressure locks for entrance and exit, everyone in white clean-room suits, and all the controls necessary to observe sterile conditions.

I reverted to my old Skunk Works axiom, "Keep it simple, stupid." The more complexity, the more potential for problems. This is too sophisticated for the Skunk Works, I decided.

We'll use the material we've worked on experimentally for ten years, the new advanced titanium in conventional structure.

The shape of the airplane itself was determined after a great many wind-tunnel and other tests. The result, head on, looks like a snake swallowing three mice. And for good reason. We added chines on the fuselage to get aerodynamic lift, among other things. We had some without them, but a terrific amount with them.

Before we got into high gear on production, we thought it advisable to build several test samples of the most complex sections: the nose and the basic wing structure.

The first wing section was a catastrophe. When we put it in a "hot box" to simulate high in-flight temperatures, it wrinkled up like an old dishrag. The solution was to divorce the skin panels from the wing spars in each direction and put corrugations and dimples in the skin—the wing surface. When the titanium got really hot, the corrugations merely deepened. I was accused of making a Mach 3 Ford Trimotor—that was made all of corrugated aluminum. But it was a very effective solution to a really difficult problem.

The nose section of the airplane presented other problems. We put it in the hot box to study cooling requirements for the pilot and the gear. We produced 6,000 parts, and of them fewer than ten percent were any good. The material was so brittle that if you dropped a piece on the floor it would shatter.

Obviously, we were doing something wrong. We queried Titanium Metals Corporation on why we had hydrogen embrittlement from our processes. They didn't know. So we threw out our entire titanium processing system and replaced it with the same methods TMC used in making the original sheets and forgings at their factory.

After the initial shambles on the nose segment heat treat tests, we put into effect a quality-control program that I believe was and is unequaled anywhere. For every ten parts manufactured, we made three sample parts. These would be heat-treated and otherwise tested before any of the others of the batch would be put in storage for future use. One sample went into a tensile strength test machine to find out how strong the material itself was. In the second, we made a short cut—about one-quarter inch long—and bent the sample at that cut around a form with a very small radius—as small as 32 times the thickness of the sheet—to see if it would crack. The third sample was used in case a re-heat-treat test was necessary. We didn't want to throw away the whole batch needlessly; it was too darned expensive.

We could trace back to the mill and know the direction of the sheet rolling, and whether the part was cut with or against the grain. Before we would do all the expensive machining to cut landing gears from the huge heavy extrusions we would cut twelve samples, and unless everyone met the test we devised for them, we would not use that extrusion to make a landing gear. We've had no landing gear failures on the birds despite the hard landings that go with in-service flying.

There were times when I thought we were doing nothing but making test samples. But the test effort was worth it. By the early 1980s, we had made more than 13,000,000 titanium production parts for all of our Skunk Works airplanes and also for the Lockheed L-1011 commercial transport and the company's big military cargo aircraft.

Titanium is such a rigid material that it cannot be shoved into place—as can some other metals—and therefore cut to less-exact tolerances. It must be tooled to fit. While this exact tooling is very expensive, it saves in the long run on scrap parts—of which there were almost none in production.

We had to invent a very large press that would shape titanium under very high temperatures—up to 1500°F and very high pressures.

The tough titanium actually is a very sensitive material to handle. Everything wants to poison it. We learned at an early date that we had to take cadmium-plated tools out of the mechanics' tool boxes if they were going to work around the engine, because the cadmium would flake off enough to poison the bolts. After only one or two runs where they attained temperatures above 600°F, the bolt heads would just fall off. We had to keep cadmium away from the titanium.

We found that the spotwelds on the wing panels failed very early in their test life when we built the panels in the summer, but if they were built in the winter they would last indefinitely. Analyzing all the processes, we discovered that in summer the water supply system for the city of Burbank was loaded with chlorine to reduce algae. When we washed the welds with pure water, there was no problem.

Special tools were required. When we first tried to drill the heat-treated B-120 titanium, a drill would be totally destroyed after about 17 rivet holes. Finally we found suitable drills developed in West Germany. Today we can drill more than 150 holes with a drill; and resharpened, it will drill another 150 holes.

We had to train thousands of people, not only our own, but Air Force mechanics and employees of our subcontractors and vendors—more than 300—in how to handle the machined parts. It's difficult to get an old-time machinist to change his ways. He wants to discover on his own how to do something. So in the Skunk Works we put them in the experimental shop under the engineers' direction and made them a party to developing the data. That always is a good tactic: involve the employee in the whole program as much as possible to arouse his interest and inspire his best performance.

One thing we learned in manufacturing the first Oxcart airplane was not to trust color coding. I had insisted on color codes for all wires and tubes and other connections, so that plumbing and other systems could not be installed incorrectly. Working with that many people, we discovered that ten percent were color blind. We've found a part bent over four inches to be connected incorrectly. We still color code, but we also use odd-shaped terminals that will fit only one way for those who can't distinguish colors.

Materials and manufacturing were only part of the problem. There were the systems—hydraulic, electrical, and others.

Redundancy of systems has been a design requirement of mine ever since we added the auxiliary fuel pump to the F-80 after we lost Milo Burcham. The Constellation had it for the first completely power-boosted controls. The Lockheed L-1011 airliner has it in triple and quadruple systems. The Blackbirds have triple redundancy. It is a safety factor that pilots especially appreciate. The cost, if you start with it in original design, is only a few tenths of a percent of the total. When you consider the value of the human lives and the vehicle this redundancy is protecting, the cost seems even smaller.

Hydraulic fluid was another special problem. First, of course, we surveyed all the suppliers to see if any of them had a high-temperature fluid, able to operate at above 600°F. One responded with literature on a fluid that worked at 960°F. I requested a sample immediately. It came in a canvas bag! That's a funny way to ship hydraulic fluid, I thought. When I opened the package I found a white crystal.

Yes, it would operate at 960°F, but it was a solid at ordinary temperatures. You'd have to thaw out the hydraulic system with a blowtorch—not too useful for an airplane. So, the hydraulic fluid became a development project, too. The final product was a basic fluid developed by Pennsylvania State University, but with seven ingredients that we added so that it would withstand the temperatures and still function as a lubricant for the pumps and other hydraulic gear.

There were a few other little items—all important. Leather washers or rubber O-rings could not be used at those temperatures. Steel was the answer, giving no trouble at high or low temperatures. Fuel tank sealants to contain the fuel were another necessary development. While the airplane is not totally tight and will leak some on the hangar floor, the fuel has such a high ignition point that it is much safer than ordinary fuel.

Electrical problems alone threatened success of the project. We were not able to complete a flight on the Oxcart without some kind of failure due to the electrical system—which controlled the autopilot, flight control system, navigation system, and with electrical transducers even the hydraulic system.

At one time, 17 percent of our flights had to be cut short because we couldn't measure oil pressure. We could not take a chance on burning out those very expensive engines. We had to institute a cooling system for the oil temperature gauge.

Especially critical was the electrical transducer measuring air displacement related to proper positioning of the inlet spike. It was our toughest electrical problem.

We simply were not able to get the electrical system to work reliably under conditions of very high altitude, very high temperature, and very substantial vibration.

I personally spent six weeks at the test base working on this problem. The entire project was at stake.

Finally, we had to invent our own wire for the electrical system. We used high-temperature Kevlar wiring and just wrapped asbestos around it in the critically hot sections.

Special plastics were designed not only to withstand the high temperatures but to provide low radar return.

The Blackbirds take their name from the dark blue-black paint. The color was determined after tests for emissivity—heat emission from the hot airplane in flight. Emissivity can make a difference of 50 to 80 degrees in temperature on the aircraft, so it is a critically important item. Actually, the color of the Blackbirds becomes blue as temperatures increase at high speed and altitude.

We had to invent a special paint for the Air Force insignia. After just one hot flight, the red would turn brown and white become mottled. Getting the paint to adhere to the plane at all was another problem. We were getting little pockmarks on the painted high-temperature plastic that makes up 20 percent of the aircraft surface. We found that when fuel had been spilled, and temperatures of 550 to 600 degrees reached, little miniature explosions were occurring on the plastic skin. The paint had to be made fuel proof as well as rain proof.

The various payloads—cameras, very sophisticated electronic gear, navigation systems, inertial systems—all were the result of great effort in development.

The inertial navigation system is so good that you can take off, put in 16 different check points, and on autopilot fly at speed, altitude, and direction desired. When the pilot says, "Home, James," the system takes him home.

The powerplant for the Blackbirds is a marvelous development on the part of Pratt & Whitney. It is the only engine of its kind in the world.

It started out as a Navy engine for an airplane that never was built. When we saw the size of it, we wanted to adapt it to high Mach numbers. P&W put in what I will call a gear shift system. When speed of a couple of thousand miles an hour is reached, the engine shifts into another cycle, bypassing the high compressor for conventional jet operation and flying as a ramjet. With no machinery obstructing airflow, the faster the plane goes the faster it wants to go.

But the hardest task of all was to make the air flow properly through the engine inlet and outlet in supersonic flight. Consider that the air hitting the tip of the spike on the inlet to the engine nacelle encounters temperatures above 800°F, then has to expand, and in the duct undergo a 50-to-one compression. This must happen without separation of airflow from the walls of the inlet duct.

It took many years to develop that inlet so the engine would operate under all flight conditions. In the early days, we often had blowouts when the flow would separate. The engine would drop from 16,000 to 20,000 pounds of thrust to about 16,000 pounds of drag—in a fraction of a second. The pilot would be so battered around by the vibration that he could not even tell which engine was out.

We solved that by putting an automatic control on the rudder, so that in .15 second the pilot could sense an engine had blown, which one it was, and kick in 9 degrees of rudder with the hydraulic steering system. The airplane then would continue in level flight.

Now we have an automatic restart. This works so well that I was concerned when the SR-71 was in service in Southeast Asia that some of the pilots flying those rough missions never had experienced a blowout; so we reintroduced practice blowouts in flight training.

Efficient as the engine is, I have kidded my friends at Pratt & Whitney that it is supplying only 17 percent of the push for the airplane in flight. Our inlet duct and the ejector make up all the rest of the thrust. Of course, we wouldn't have the rest of it without that vital 17 percent!

When President Johnson first announced the A-11, the photo released actually showed its successor, the YF-12A, in flight. In developing the triple-sonic planes, we worked through design numbers A-1 to A-12. A-11 was just a design number. We built 12 of the A-12s. The YF-12 photo probably was substituted for security reasons. It revealed less about aircraft capability than another model.

The YF-12 first flew on August 7, 1963. The Air Force had had difficulties with an earlier candidate for a high-altitude interceptor and accepted our proposal to make a long-range, high-speed, high-altitude interceptor. Gen. Hal Estes wanted us to incorporate the ASG-18 radar system that Hughes Aircraft had been working on and to be able to fire the Hughes GAR-9 missile. We used both, but a substantial amount of development testing still remained.

With its large fuselage, the YF-12 could easily contain three of the large missiles. But no one ever had fired a missile from such high altitude and at such high speed. It took us three years to do it.

It was a real development problem to be able to open the bomb bay doors, eject the missile, and ignite it, so that it would hold its path and not, instead, come up between the first and second cockpits. There had been big problems with launching missiles from aircraft even at much lower speeds.

To keep the missiles away from the airplane, we had to develop thrusters that would push down with a certain force on the forward end of the missile and another force on the rear—all in just a few seconds. The missile was ejected to a distance about 40 feet below the airplane and then fired.

Launched from the airplane at well above Mach 3, and accelerating on its own at another Mach 4, the missile was speeding hypersonically at Mach 7 at the peak of its course. We fired at target drones at altitudes ranging from sea level to more than 100,000 feet, and hit targets more than 140 miles away. We proved that we could hit targets flying over ocean or over land.

Our success rate was better than 90 percent hits. The GAR-9 missile and the ASG-18 radar system performed beautifully. Development of the missile in the YF-12 led to the Phoenix system used in the F-14 today.

Three YF-12As were built on the prototype contract, and from their performance the Air Force determined that a fleet should be produced for the Air Defense Command. Gen. Arthur Agan needed replacements for the outmoded F-102s and F106s. At one point, he said publicly that with his inadequate interceptors and without a radar network he could not protect Air Force One flying the President across this country from Washington to Los Angeles.

Three times in three years Congress appropriated funds, more than $90 million, to start production of 93 of the big fighters—F-12B—for the ADC. But Secretary of Defense Robert McNamara, guided by his "Whiz Kids," decided no need existed for such a high-performance airplane, that our potential enemies did not have anything comparable, and that an airplane couldn't hit a target going that fast anyway. Of course, the airplane had a better than 90 percent hit rate in firings at Mach 3.

And the threat that wasn't there soon surfaced in the Russian supersonic Backfire bomber.

The Backfire can outrun any of the American fighters, including the F-15, and F-16, and others. For that matter, so can the commercial Concorde transport; both Concorde and Backfire are in the Mach 2 category. Our present fighters have a supersonic range of only 50 to 100 miles before they exhaust their fuel supply. The YF-12A, now principally assigned to NASA research, has the speed and range at Mach 3-plus. The F-12B, of course, could easily have flown up to inspect or shoot down a Backfire, which, while capable of Mach 2, can reach the speed only for brief runs, not sustained cruise.

Rather than re-equip the Air Defense Command, the decision then was made to downgrade its role.

Even without a contract for production of the plane, the

existence of F-12B tooling was viewed by the McNamara team as a threat to funds for the B-70 supersonic bomber and the F-15 fighter. We retained the tooling for three years, but then were ordered to scrap it—which we did at 7½ cents per pound. The cost to replace it today and put such a plane into production would be totally out of the ballpark—hundreds of millions of dollars for the tooling and $70 or $80 million for each airplane. At the time, the airplanes would have cost about $19 million, and there is no fighter-interceptor in this country's air forces today with comparable capability.

But the Blackbird technology was to see other applications. The Air Force wanted an advanced reconnaissance airplane that could carry larger cameras and more sophisticated gear than the glider-like U-2. The customer was the Strategic Air Command, which in my book is the most sophisticated of them all for development of aircraft and materiel. The SR-71 was the result.

By the end of 1962, we had an initial contract to build six of them. I should say the RS-71 was the result, but when President Johnson announced first flight of the airplane two years later in 1964, he reversed the initials. It meant a lot of scrambling by SAC and the Skunk Works to revise the designations. The original meant "reconnaissance strike;" the revision was called "strategic reconnaissance." We still use the nickname "Strike Recci."

The SR-71 carries a pilot in the front cockpit and a recon-naissance officer in the rear to handle the cameras and certain navigation jobs. It is a much heavier airplane than the earlier Blackbirds in that it carries more fuel and more and improved instrumentation. It looks much like the earlier models, but the structure is considerably improved. And by the time we got to the third Blackbird we had that air inlet development well in hand, so the flight test program went ahead rapidly.

It was on the SR-71 that we worked with Wyman-Gordon to develop large titanium forgings that we could machine down to final form for installation on the aircraft. It took $1,000,000 for Wyman-Gordon to develop the forging methods, and it cost

Two other aeronautical masterpieces in which Kelly Johnson had the essential role as a legendary contributor to national security: the SR-71 and TR-1.

the Skunk Works another $1,000,000 to learn how to cut the forgings into parts. But, for example, one forged part replaced some 96 parts that we had had to build up into a single part for prior airplanes. The net result was a saving to the military, and the taxpayers, of $19.5 million from that single program.

It has become something of a crusade with me to advocate development in this country of a huge metal forming press. We need a 250,000-ton machine, five times larger than the biggest one available to us here today. When we have to machine away 90 percent of rough forgings to make very large aircraft parts—specifically, titanium nacelle rings and landing gear for the SR-71 and aluminum fuselage side rings for the C-5 transport—that seems very wasteful. The Russians have not hesitated to make this investment. They have more and larger forging presses than the United States.

In-flight refueling, developed to give the SR-71 global range, has become routine. By the early 1980s we had made more than 18,000 of them and can refuel in the air anywhere.

In fact, I have suggested that mid-air refueling could allevi-
ate the sonic-boom problem for supersonic transports. We've
had to be very careful where we flew with the Blackbirds
because of that problem—shock waves carried to earth in the
wake of the aircraft. Inevitably we've heard complaints on sonic
booms—everything from claims that they disturbed fishing in
Yellowstone Park to one insistence that they made pack mules
want to jump trail. I had a complaint, too, when one of my
military friends "boomed" my ranch and broke a $450 plate-
glass window. Naturally, I got no sympathy from anyone on
that.

The SR-71 has been in service with SAC since 1965. It is the
fastest and highest flying airplane in service anywhere in the
world that we know of. The Blackbirds have a background of
experience in flying faster than Mach 3 unmatched by any
aircraft in any country.

On April 27, 1971, a USAF SR-71 cruising above 80,000 feet
set duration and distance records on a 10½ hour flight covering
15,000 miles for which the Air Force received the Mackay Trophy "for the most meritorious flight of the year," and the
Harmon International Trophy in 1972 for "the most outstanding international achievement in the art/science of aeronautics."

On September 1, 1974, a USAF SR-71 set a transatlantic
record of 1 hour, 54 minutes, 57 seconds, on the 3,470 statute
mile course from New York to London, en route to the semian-
nual Farnborough Air Show and its first public appearance
before an international audience. The return flight from London to Los Angeles established a world speed record for the
distance—3 hours, 47 minutes, 35 seconds, over the 5,463
statute miles.

On July 27, 1976, Strategic Air Command pilots flying out
of Beale AFB in California set six new world records for speed
and altitude, including closed circuit speed of 2,086 mph;
straight course of 2,189 mph; and absolute altitude and altitude
in horizontal flight of 86,000 feet.

There is a fourth Blackbird, and until very recently I was

not allowed to talk about it. As so often happens, clearance came because of a news photo and story. A number of the aircraft were shown in mothballs at Davis Monthan AFB in Arizona.

This is the highest performing of all the Blackbird series—it flew higher, faster, and farther. It is an unmanned remotely piloted vehicle—the D-21, a drone. As our equipment becomes smarter, more powerful, more sophisticated, I wonder about the need always for a man in it. The answer may be to leave the man on the ground, but in control of the vehicle.

15

In Sickness and in Health

WHEN WATER CAME TO OUR LINDERO RANCH from the Colorado River aquaduct, taxes rose 1,000 percent. The property was taxed then not as ranch property but at its best use, which would be for subdivision. It was evident that we couldn't continue to run the ranch with taxes of that magnitude. We had no choice but to sell it, which we did in 1962.

The developer initially built 746 homes on the ranch after we sold it. He dug a lake near the Ventura Freeway and called the new community Lake Lindero.

Althea and I already knew well the area where we wanted to buy our next ranch. It was in Santa Barbara County, near Alisal Ranch where we had visited frequently, played golf, and ridden the trails over 10,000 acres. I scouted the territory from a small plane, flying over the mountains and valleys of this beautiful unspoiled country. In fact, I learned to fly just for this purpose.

Star Lane was exactly what we wanted, and we bought it in 1963. The name came with the ranch, and I especially wanted to retain it because it is in the Lockheed tradition of naming our airplanes and now our space vehicles after stellar bodies—Vega, Orion, Sirius, Constellation, Shooting Star, Starfighter, and the Polaris, TriStar, and Galaxy of today. It is a working ranch 30 miles north of the city of Santa Barbara, northeast of the picturesque Danish community of Solvang in the Santa Ynez Valley—nearly 2,000 acres, on which we run 300 head of cattle and raise oat hay. It lies in a valley about three and a half miles long and about a mile wide, totally isolated from other ranches.

The former owner was a motion picture distributor who had built a poolhouse complete with motion picture projection equipment, as well as a main residence. The main house is beautiful in the Spanish style and is somewhat larger, in fact, than our Encino home. And the pool is ten feet longer and five feet wider than the Encino pool. We weren't exactly roughing it there.

This ranch and, I think, most ranches in the area are covered by California's Williamson Act, which protects truly agricultural property from the untenable taxation that drove us from Lindero. Of course, this imposes a responsibility on the owner, too. There are restrictions on dividing the property; and should it be subdivided for other than agricultural purposes, taxes would be imposed retroactively at the highest use rate.

So, we didn't have to build our home on the new ranch, just move our equipment from Lindero. It still was a do-it-yourself operation. We bought a big new truck with a 21-foot flatbed for hauling our tractors and other equipment. All the machinery necessary for a working ranch requires a lot of maintenance, so one of my first projects was to build a shop. I had a lot of fun designing it. Having a shop is also one of my joys, and has been since boyhood. I respect machinery and want it taken care of properly. This I do myself on the ranch, saving a great deal in maintenance costs.

My present shop is a large one—40 feet wide, 120 feet long, with a 22-foot-high gable roof. The roof is stressed to take loads of 6,000 pounds so that I can hoist heavy machine elements when necessary. The doors are designed to tolerate wind velocities up to 140 miles an hour, and the building is stressed for earthquake loads of one-quarter unit of gravity laterally.

It is well equipped with lathes, grinders, power saws, welding equipment, a hoist for engine repairs—everything needed to keep all the farm machinery running. It also houses the farm's working vehicles—six tractors, four trucks, the hay baler, and other equipment.

We built a new house for the ranch foreman and his family and refurbished the original ranch house and a second small frame house for other ranch hands.

We moved our windmill from Lindero—took it all apart and reassembled it—and installed three new ones. They're all painted yellow like daisies. I tell first-time visitors that they're cow fans to keep the cows cool in summer. We have seven wells and plenty of water on the ranch but cannot count on the wind for 100 percent of our electric power. Gasoline-powered pumps in the field provide water for the cattle, and we depend on the utility company for the all-electric heating in the main house.

Star Lane was everything Althea and I had wanted our ranch to be. Again she was a full partner with me in its operation. She loved the outdoor life as much as I did. And for me, of course, it was a life-saving escape from the pressures of work.

Although not all that safe, either! There was one time when I had interrupted my work riding the tractor and discing a field to take the jeep and pick up our foreman, Lee Erickson. As we were driving along we saw a little calf lying in the middle of the road. We thought he might be injured or sick, so we got out of the jeep to check.

Now, the one thing you do not do is touch a young calf if the mother is anywhere nearby and you want to stay alive. But Lee gave this one a gentle boot in the tail to see if it could move. Well, it could and did, but it also bawled loudly. I decided to let Lee go on up to his house alone and return to my tractor to finish the discing job.

I hadn't walked more than a hundred feet when suddenly there was a tremendous noise behind me, a great roar, and charging at what must have been at least 140 knots was what looked like the biggest cow I'd ever seen—the calf's mother, of course. She should have charged Lee, but he was safely in the jeep, and I was the nearest target.

Thank heaven, I had disced that part of the field pretty thoroughly. I got down in the loose dirt as quickly as I could, falling on my back; there wasn't time even to turn and fall face down. The cow went right over me, udders dragging over my face, but hooves not hitting with full force fatally, I hoped. Fortunately, the cow didn't return to finish the attack on me but rejoined her calf and led it back into the hills.

When I discovered that I could get up, I wondered if I had any broken bones, but decided to try to finish the discing. I figured that I'd feel it soon enough if I had. I was able to finish the job, and when I went up the hill to join Althea at the house for lunch she couldn't believe what she saw.

"What on earth happened to you?" she asked. My clothes were torn and I was black and blue. There were big bruises on my rib cage where I'd been stepped on—but nothing was broken, apparently.

What an ignominious end it would have been if I'd been killed by a cow after all those thousands of hours of flying in experimental airplanes! I've always said it was safer in the air than on the ground.

When I told the story back at the Skunk Works, one of our engineers of Spanish descent and knowledgeable in the art of bullfighting gave me a toreador's red cape and demonstrated how to use it should I have occasion again to have to duck a charging cow.

For about two years after we bought Star Lane, Althea and I lived an almost idyllic life, not without its pains and difficulties, but overall a very happy time.

My ulcer problem, which I'd fought since my college days, recurred seriously during the compressibility problems with the P-38. One evening during dinner in our Encino home, I just fell off my chair onto the floor. I had been suffering from ulcer symptoms again, and I knew this time it had burst—although I'd never had that experience before.

The pain was severe, and Althea's first reaction was to give me a stiff double shot of brandy. Not exactly the conventional treatment, but it eased the pain. I had a lot of work to do at the time, so I went to the plant next morning and didn't get around to reporting to the company doctor for two or three days, even though I knew the ulcer was still bleeding.

"I've had some trouble," I told Dr. Lowell Ford.

"You bet you have. You're crazy," he told me when he heard I'd gone to work. But I also had healed in that time. Despite the fact that it usually would not have been prescribed for ulcers, forbidden, even, the alcohol apparently had a relaxing effect on the tensions that led to the ulcer formation. Eventually I was able even to persuade the doctor, who also had ulcers, of the validity of this treatment—in moderation.

During the war years, I was working on as many as six airplanes at one time and making a trip once or twice a month to Washington. In the pre-war DC-3 transports, that meant stops across country; flying was not as enjoyable as it is today in the big jets. At the plant, I would start about six o'clock in the morning and maybe work on the F-80 for an hour or so. Then I'd move on to the P-38 problems while these still were outstanding. Various derivatives of the Hudson were also in development—the Ventura and later the PV-1. The Constellation we kept going, too. And we had responsibility for the turbojet engine development headed by Nate Price. There were enough problems to keep me busy.

My career had moved ahead. After being chief research engineer from 1938, I was made chief engineer in 1952.

With the job came more responsibility and more ulcers. I had had more fun as chief research engineer responsible for designing and testing airplanes. As chief engineer, I had to learn quickly about translating the engineering into production. I discovered I had inherited a department of 5,500 people, about 800 of whom were engineers, and a complex system of drafting, printing, issuing, controlling, and revising production drawings. A simpler system with fewer people was the way I preferred to go, but I couldn't use the Skunk Works approach with an army of 5,500.

Again I was trying to do many jobs at once. There were later versions of the Constellation transport. The F-104 and U-2 had come along. In those days, whenever one of our aircraft had an accident, particularly a fatal accident, I would develop a stomach ulcer in about 24 hours. One developed over that fatal night operational test with the F-80 and B-25 when they collided. Another developed over the Constellation grounding. The solution to that when discovered was so simple it took about 15 minutes to re-engineer. I found the darned part myself—a fitting where the aluminum wiring was all scorched. The insulation had been inadequate. We had trouble with the engine backfiring, and we added fuel injection to counter that.

Fortunately, the worrisome nature that produced immediate ulcers during a crisis was overridden by my naturally strong Swedish constitution, and the ulcers would heal in about a week.

All of this took its toll on me physically, though, and on March 25, 1955, an interdepartmental communication to company supervision announced, "On the advice of his doctors . . . Johnson will take a complete rest . . . (his) health has suffered from the unusually heavy work load pressures his position has forced him to carry during the past many months . . . (On) return to work, his activities will be on a reduced basis and limited for an extended period of time to top level technical assignments."

By 1956, I had been made vice president for research and development. This meant the added job of visiting various divisions of the corporation—the military aircraft production company in Georgia, for example, and the missiles and space division in northern California—to coordinate research and development and to avoid duplication of effort. This was a job I did not like because I did not have direct authority.

This business of suggesting is not my modus operandi. I much prefer to have direct authority, as I have in the Skunk Works, where there is no argument about what will be done nor how to do it. I was, therefore, in my own opinion, a very poor corporate engineer. By 1959, I had decided and announced, "I want no more of this. I don't like all this traveling. I don't like this job. I'll put in full time at the Skunk Works."

By that time, since 1958, a new title had been created for me as Vice President for Advanced Development Projects. That was the Skunk Works. My stand was accepted, and darned good timing it was, too. It was a very busy period and about that time we started work on the Oxcart. That would take all the energy I could put into it. And give me another ulcer.

Taking X-rays during an annual physical examination about 30 years after that first time I collapsed with a perforated ulcer, Dr. Ford said, "Kelly, the outlet to your stomach is about as big as a lead pencil. If you don't do something about it you will die. It could happen in a few weeks!"

At that point, I had accumulated so much scar tissue that I no longer felt any pain in the area. So, in 1970, I submitted to an operation that removed half of my stomach. Fortunately, duodenal ulcers tend to recur in the same place, so I still have a functioning stomach and have not had another ulcer since. It was a good weight reduction program, too, costing about $500 a pound, approximately what it cost on the C-5 cargo airplane.

Through all of this, our ranch—first Lindero then Star Lane—was a welcome retreat. In 1964, a year after we made the move to Star Lane, I was elected to the Board of Directors of the Lockheed corporation. By the next year, Althea began to feel less strong than her usual buoyant self. A physical examination showed that she had cancer.

It was a devastating blow to both of us. After two operations, we both knew that she would not overcome the deadly disease. She suffered a severe depression during which she attempted to put herself to sleep permanently with pills so that she would not be a burden to me. Fortunately, this happened at the ranch on a day when I returned earlier than expected from working in the fields. I was able to rush her to the hospital, and she kept going for another few years.

I've often wondered if a serious automobile accident we had several years earlier, when a drunken driver hit our car broadside, causing head injuries to Althea, had anything to do with her later difficulties.

It was a very, very trying period. Our good friend, Dr. Ford, spent a great deal of time with us, staying overnight at the house after having worked a full day. He not only remained for Althea's sake but to monitor my own heart. I was having angina attacks during the night.

One of Althea's hopes had been to endow a chair in my name at Cal Tech, and this she did. It is funded to begin after my death. Althea endured three more operations before she died in December 1969.

She had not wanted to be put in a grave, but rather to be cremated and have her ashes scattered over Star Lane, the ranch she loved so much. This was not legal, we discovered. So, with Dr. Ford and Tony LeVier, I piloted a small airplane over the mountains where we had ridden so frequently in happier times, out over Santa Barbara bay, and far enough out to sea where legally we could fulfill her last wish.

16

It's No Secret

WHAT THE SKUNK WORKS DOES IS SECRET. How it does it is not.

I have been trying to convince others to use our principles and practices for years. The basic concept as well as specific rules have been provided many times. Very seldom has the formula been followed. One exception was the successful development of the Agena-D space vehicle at Lockheed Missiles and Space Company. Another was the Army's management of its Division Air Defense Gun Program.

But I fear that the way I like to design and build airplanes one day may no longer be possible. It may be impossible even for the Skunk Works to operate according to its proven rules at some point in the future. I see the strong authority that is absolutely essential to this kind of operation slowly being eroded by committee and conference control from within and without.

The ability to make immediate decisions and put them into rapid effect is basic to our successful operation. Working with a limited number of especially capable and responsible people is another requirement. Reducing reports and other paperwork to a minimum, and including the entire force in the project, stage by stage, for an overall high morale are other basics. With small groups of good people you can work quickly and keep close control over every aspect of the project.

The lesson I learned early from Hall Hibbard about not

Skunk Works insignia.

driving people has served the Skunk Works well. People challenged to perform at their best will do so. With rare exceptions, long hours are not encouraged.

"If you can't do it with brainpower, you can't do it with manpower—overtime," is axiomatic with me.

Our aim is to get results cheaper, sooner, and better through application of common sense to tough problems. If it works, don't fix it.

"Keep it simple, stupid"—KISS—is our constant reminder.

"Be quick, be quiet, be on time," is another of our mottos.

"Listen; you'll never learn anything by talking. The mesure of an intelligent person is the ability to change his mind."

These concepts save time, money, and people.

The Skunk Works at Lockheed has moved four times since the first shop was constructed of engine boxes and a tent in 1943, and its first project, the XP-80 jet fighter, was built with just 120 people in 143 days. There were only 23 engineers on the project. There were 37 engineers on the JetStar corporate transport. The U-2 many years later employed a total of 50 people on both experimental and production engineering. On the enormously more difficult SR-71, there were only 135 engineers.

The present Skunk Works No. 5 was christened by Althea in January 1963. By that time, the organization had turned out 17 major projects and participated in two others.

Another secret of the Skunk Works that is no secret is our human relations. The Skunk Works never has had any serious labor problems. We've had fine relations with the union. The union president would listen and be responsive when we told him our problems. His stewards would, too. At the test base in Nevada, I would meet with union stewards to hear their problems—which were real and many—and always try to do something about them.

On one occasion when Lockheed was threatened with a strike, Tom McNett, then president of IAM Lodge 727, told me, "Kelly, of course, we won't strike the Skunk Works."

There was a strike, briefly. The Lockheed plant as a whole historically has had excellent labor relations. The company was used, in fact, as the example in a government booklet, "Causes of Industrial Peace," published in the early '40s.

While the union had to make a token protest at the Skunk Works, pickets were stationed off to the side of our main gate. They let us operate.

I personally have had perhaps 20 or 30 union grievances filed against me over the years for performing work of some kind that a mechanic should have done. An example was welding a door fitting for the JetStar in my shop at home. But this has all been without hard feelings. I've always thought the employees actually enjoyed the fact that their top boss cared enough to work right along with them.

One of my challenges to employees is a standing twenty-five cent bet against anyone who wants to differ with me on anything. I keep a supply of quarters on hand. It's not the quarter, of course, but the distinction of winning it—to be able to beat the boss. It's another incentive. And I've lost a few quarters, too.

It is our practice to put the people in close contact with the airplane while it is being built so that they can follow it throughout its development. They feel responsible for the parts they make. If a part needs fixing, it will be fixed quickly.

On engineering changes, we go out of our way to explain them to all who will be working on them. We maintain a very close liaison from me to the designer, to the purchasing agent—so that he understands the urgency in acquiring materials; to the tooling people; and to the people who actually will build any part of the plane. That carries through from the first line I draw on paper to completion of the airplane and its first flight.

It long has been my practice to see that all those closely connected with a project witness the first flight. This dates from the days of our first Skunk Works airplane, the XP-80. We bussed our crew to the desert test base for the flight and had a party afterwards, with the Indian wrestling matches that became a sort of tradition.

Involving families of employees whenever possible is important, too. When we christened the present Skunk Works, we held an evening party in the new plant for all employees and their families. We told them as much as we could about our work. The next day we went back to strict security with no admittance without need.

Security regulations necessarily are impersonal and unyielding. An excellent engineer, a fine and honorable person, can marry someone who might have a relative with political views unacceptable to our government, for example, and thereby become unemployable in a secret operation. It was painful, but I have had to say goodbye to a good friend for such a reason.

Most companies, while desiring the benefits, will not pay the price in revised methods and procedures for setting up a Skunk Works-type of operation. They will not delegate the authority to one individual, as Lockheed did in my case from the very first Skunk Works. It requires management confidence and considerable courage.

Without the authority assigned to the Skunk Works by our

military customers and the Lockheed corporation, we would not have been able to accomplish many of the things we have done, things about which I felt we could take a risk—and did.

The theory of the Skunk Works is to learn how to do things quickly and cheaply and to tailor the systems to the degree of risk. There is no one good way to build all airplanes.

I believe that the designer and builder of an airplane also should test it. That is important to his ability to design future aircraft. I always have thought of flight testing as a method of inspection—to see how well you have engineered and built the bird. Working with pilots in testing and development has taught me a great deal over the years. Pilots are a special breed for whom I have a deep admiration and respect. When the day comes, if ever, when we do not have responsibility and authority to test the airplanes as we design them, then from that day on our design ability diminishes. We will lose competency to develop new aircraft.

There is a tendency today, which I hate to see, toward design by committee—reviews and recommendations, conferences and consultations, by those not directly doing the job. Nothing very stupid will result, but nothing brilliant either. And it's in the brilliant concept that a major advance is achieved.

Development of some of this country's most spectacular projects—the atom bomb, the Sidewinder missile, the nuclear-powered submarine—all were accomplished by methods other than the conventional way of doing business outside the system.

Operating at its best on our Air Force programs, the Skunk Works could get an almost immediate decision on any problem. I could telephone Wright Field, Dayton, for example, talk to my counterpart who headed the small project office there for the military—and who was allowed to stay with it to conclusion—and get a decision that same morning. Now, that's just not possible in standard operating procedure. It's a difficult concept to sell for the first time, though, since it means abandoning the system.

As well as giving us the authority we needed, Lockheed also gave us the tools. The importance of a research capability is basic to advanced engineering design. Lockheed management from its earliest days has been far-seeing in this regard. The company not only built the first privately-owned wind tunnel in the industry—back when we were working on the P-38—but today in its Rye Canyon Research Center operates the most complete advanced research and test facilities integrated at one site of any aerospace company. (It was renamed the Kelly Johnson Research and Development Center in 1983.)

In 1954, when I was chief engineer, I was able to get Messrs. Gross, Chappellet, and Hibbard to commit $100,000 for study of such a center. Requirements outlined at the time included test tunnels of supersonic capabilities. Today, there are hypersonic, hypervelocity, and propulsion tunnels, space chambers, laboratories in electromagnetics, cryogenics, acoustics, thermal systems. The ability to "fly" an airplane on the ground and through several simulated lifetimes of service using various mockups has become a science in itself. Long before an airplane ever is assembled for flight testing, we now know quite precisely how it will fly, what kind of maneuvers it can perform, the shortcomings of its systems, and any elements that may be likely to fail. Without these excellent research tools, the Skunk Works would not have been able to build and fly the advanced aircraft it has produced.

Before the decision was reached to locate the new research center on several hundred acres in Rye Canyon—a relatively remote and ruggedly beautiful setting in the foothills of the San Gabriel Mountains—we investigated other potential sites.

The final selection of Rye Canyon was influenced by its convenience to the Burbank and Palmdale plants, being about equidistant to both. And the surrounding foothills would provide privacy and act as sound barriers to contain the noise from our tunnels and other test facilities. An original 200 acres purchased in 1958 has been expanded to some 500 today. This research headquarters is entirely company-funded. The laboratories perform work under contract to other divisions of Lockheed—as well as other companies and government agencies when time is available.

At the Skunk Works, a typical day for me would start about 7 a.m., except in a crisis situation when I would arrive at six o'clock to make up for the three hours' time difference with military offices in Washington. Generally, I would hold a meeting with the engineers working on the key problems. Almost always the group included my three top assistants—we're short on titles in the Skunk Works: Dick Boehme, Rus Daniell, and Ben Rich, my successor as head of ADP today. Ben is one of the few persons to win one of my quarters. The bet was over how many degrees' temperature change painting the Oxcart would mean. I guessed it would lower inflight skin temperature 25 degrees. Ben, a thermodynamicist, guaranteed 50 degrees, and he was right. It was 52 degrees.

These morning meetings were short and informal. I liked to sketch on a pad of unlined yellow paper the specific work we should do and outline the program for the day or week. The pads of lined paper that most engineers use I find inhibiting. It was nine years before I discovered that my loyal secretary, Verna Palm, was having these pads of unlined yellow paper made up especially for me. I had thought I was being frugal and using inexpensive scratch paper. Verna was my first secretary when I became chief research engineer. Hibbard, with typical generosity, let her leave his office and work for me because he knew it would help me in my new position to continue to work with the secretary we had shared. She stayed with me for 18 years, until her retirement.

When Skunk Works principles really are applied, they work. An example of their successful application was development of the Agena-D launch vehicle. The Rand Corporation has recorded it in a report available to anyone interested.

The satellite that was to become this country's workhorse in space was in trouble in terms of design and cost but especially in reliability, which stood at an incredibly low 13.6 percent. I was drafted, in effect, to go up to the Lockheed Missiles and Space Company and fix it. We set up a Skunk Works operation with the company's design project engineer, Fred O'Green, as head. Air Force Col. "Hank" Cushman was the customer's counterpart.

O'Green's excellent performance later attracted the attention of Litton Industries, where he became president, and then chairman of the board. Cushman was promoted to general and put in charge of the Air Force's Armament Division at Eglin Field, Fla.

It proved again our axiom: If you have a good man and let him go, he'll really perform. In terms of today's world, that axiom should apply to women as well.

When I first reviewed the Agena project, I discovered that 1,206 people were employed in quality control alone, achieving only that 13 percent reliability factor! It should have been the world's most reliable vehicle just using the inspection department. That was enough people to design and build the thing.

At the Baird Atomic Company, which made the vehicle's horizon sensor, Lockheed had 40 people inspecting, coordinating, and reporting. Yet Baird had only 35 people building the instrument. We resolved that situation by returning responsibility for the product to the vendor. For example, I telephoned Walter Baird personally since he and I had worked together on a number of other Skunk Works projects. He immediately agreed to pick up his end of the log.

The same understanding was reached with the other vendors. It's a basic principle of delegating authority. Suppliers and others associated with a project must be extended the same kind of rules and permissions that are given us for the entire program. This cuts red tape and costs and allows all participants to concentrate on the product instead of a system. It is so simple.

Other measures taken saved the government at least $50 million on costs. This application of Skunk Works methods completed in nine months what had been scheduled for 18. Some 350 drawings were created against a projected 3,900, and quality control personnel were slashed to 69. Tooling costs were reduced from a projected $2,000,000 to $150,000, and procedures were established to turn out design drawings in a single day instead of one month. These measures demonstrated results. In the first twelve launches, reliability was 96.2 percent.

Attempts to apply Skunk Works techniques, however, have not always met with success. One example was the U.S. Army's Cheyenne rigid-rotor helicopter program.

The Army became interested in the Skunk Works approach because it promised quicker deliveries, greater flexibility, and lower costs than a conventional operation.

In preparation, I took Jack Real, a very able engineer and manager who was to be in charge of the program for Lockheed, plus six of his top supervisors, into the Skunk Works for six months' study of our operation.

At Van Nuys Airport, they had at their disposal a large hangar, well-lighted drafting rooms, and anything else they might need. I imposed on Real the requirement that he try to design the helicopter so that it could be serviced with six simple tools—any six of his choice. This was more a challenge than an arbitrary decision. I think most good designers want to keep things simple, but sometimes, for the sheer engineering delight of creating, things become unnecessarily complex and cumbersome.

The rigid rotor concept, pioneered by Lockheed's Irv Culver and Frank Johnson, was much simpler and safer than conventionally designed helicopter propulsion systems and had been proven successful on smaller-scale flying machines. The Cheyenne AH-56A would be its first application to a large military vehicle. The Cheyenne was designed for high performance, maneuverability, evasive operation, and was to be ideal for nap-of-the-earth flying.

Real and his team began with great enthusiasm to apply our operating methods to meet the Army's design specifications. But within six months, the satellite Skunk Works had a purchasing department larger than my entire engineering department working on seven projects. They had become buried in the usual paperwork already.

Despite the best of intentions, the Army had at the time ten different test centers and bases involved in the procurement of new weapon systems. And when you have that many representatives involved in design and development, with no single person in charge to represent the customer, the Skunk Works concept cannot work.

It is absolutely imperative that the customer have a small, highly-concentrated project office as a counterpart to the Skunk Works manager and his team. It is not a concept easily adopted after years of working within the system. There has to be an all-out commitment, or the method will not work.

The Cheyenne program was cancelled, I think unwisely, when it encountered a rotor problem. We lost a test vehicle when a rotor shed its parts. We were able to determine the cause—it was in the whirl mode—and knew how to fix it. But the Army decided to cancel the program and start from scratch.

For the money later spent in development of a helicopter with lesser capabilities, the service could have had some 450 Cheyennes. At the time the Cheyenne contract was cancelled, 145 Army personnel were involved in the program. In contrast, the total at the Skunk Works for both CIA and Air Force representatives in our U-2 and SR-71 programs did not exceed six people.

I am convinced of the military's intention to improve weapon systems development through faster and cheaper means. Taking on a major new project requires bold decisions. It is extremely difficult to predict technology problems five to ten years ahead and commit to solutions.

The Army did decide again to work by Skunk Works rules, and this time they made it work. In initial planning for development of the Division Air Defense gun and radar, I was asked for and gave a personal briefing, extended to about six companies competing as suppliers for the project. Representatives from the companies spent several days at the Skunk Works. They had been informed by the Army that they were to prepare their bids for a project of potentially several billion dollars on the basis of our "14 Points."

The basic operating rules of the Skunk Works are:

1. The Skunk Works manager must be delegated practically complete control of his program in all aspects. He should report to a division president or higher.

2. Strong *but small* project offices must be provided both by the military and industry.

3. The number of people having any connection with the project must be restricted in an almost vicious manner. Use a small number of good people (10 percent to 25 percent compared to the so-called normal systems).

4. A very simple drawing and drawing release system with great flexibility for making changes must be provided.

5. There must be a minimum number of reports required, but *important* work must be recorded thoroughly.

6. There must be a monthly cost review covering not only what has been spent and committed but also projected costs to the conclusion of the program. Don't have the books ninety days late and don't surprise the customer with sudden overruns.

7. The contractor must be delegated and must assume more than *normal* responsibility to get good vendor bids for subcontract work on the project. Commercial bid procedures are very often better than military ones.

8. The inspection system as currently used by ADP, which has been approved by both the Air Force and Navy, meets the intent of existing military requirements and should be used on new projects. Push more basic inspection responsibility back to subcontractors and vendors. Don't duplicate so much inspection.

9. The contractor *must* be delegated the authority to test his final product in flight. He can and must test it in the initial stages. If he doesn't, he rapidly loses his competency to design other vehicles.

10. The specifications applying to the hardware must be agreed to in *advance* of contracting. The ADP practice of having a specification section stating clearly which important military specification items will not knowingly be

complied with and reasons therefore is highly recommended.

11. Funding a program must be *timely* so that the contractor doesn't have to keep running to the bank to support government projects.

12. There must be a mutual trust between the military project organization and the contractor, with very close cooperation and liaison on a day-to-day basis. This cuts down misunderstanding and correspondence to an absolute minimum.

13. Access by outsiders to the project and its personnel must be strictly controlled by appropriate security measures.

14. Because only a few people will be used in engineering and most other areas, ways must be provided to reward good performance by *pay not based on the number of personnel supervised*.

My early definition of the Skunk Works holds true today:

"The Skunk Works is a concentration of a few good people solving problems far in advance—and at a fraction of the cost—of other groups in the aircraft industry by applying the simplest, most straightforward methods possible to develop and produce new projects. All it is really is the application of common sense to some pretty tough problems."

My promise then to Skunk Works employees still applies now:

"I owe you a challenging, worthwhile job, providing stable employment, fair pay, a chance to advance, and an opportunity to contribute to our nation's first line of defense. I owe you good management and sound projects to work on, good equipment to work with and good work areas. . . ."

Our employees could tell, I think, that I really believed in the Skunk Works and in them. The bottom line is integrity, and I've never built a plane in which I did not believe. Examples: The liquid-hydrogen design already mentioned; a nuclear-powered plane in the '50s; and an experimental vertical-rising aircraft, the XFV-1, which we advised the Navy was so underpowered with the engines available at the time, in the '50s, that

The vertical ascent XFV-1, an unsuccessful design concept. It was characteristic of Kelly Johnson to write off aircraft developments he came to view as impractical—regardless of the financial and manhour loss of initial investment.

it was dangerous. The Navy agreed to abandon the development effort.

Three times I was offered a company presidency at Lockheed and three times declined it. To me, there was no better job within the corporation than head of Advanced Development Projects—the Skunk Works. I was doing what I'd wanted to do since I was 12 years old.

17

Farewell, Sweetheart

BEFORE ALTHEA DIED, SHE URGED ME TO REMARRY. She did not want me to be lonely. I endured it for awhile—fortunately this was a period of demanding work for me—but I knew I could not continue to live that way. Much as I love my work and always have—perhaps more than most people—I also believe life should be shared to be really meaningful and be balanced with the pure pleasure of recreation.

My secretary at the time was a pretty, petite redhead. She had worked for me for ten years, the last two actually as an administrative assistant. She not only was beautiful but well educated and talented, a former ballet student. She became more and more important to me personally; and in May of 1971 I married Maryellen Elberta Meade in the little Lutheran church at Solvang, near the ranch. We honeymooned in Hawaii.

Once again I was happy in my personal life as well as in my profession. I knew that Maryellen was, too. She was my "Sweetheart." But not for long, about a year and a half.

In the brief, happy period we enjoyed, Maryellen came to love Star Lane as I did. We rode horseback together over the ranch whenever we could. She took up golf, too, playing at Lakeside Golf Club in Toluca Lake, where I have been a member for many years. As a relative beginner, she teamed frequently with another player of comparable handicap, Nancy Powers Horrigan.

But the diabetes that had been diagnosed in Maryellen

some years earlier but held under control became more serious. The first serious effect was to her eyesight. We tried everything. She underwent hundreds of laser treatments to reduce the blood spots on her retina, but the treatments were only slightly successful.

An operation developed at Stanford University offered hope. She submitted to the procedure twice, the second time knowing that the entire eyeball might have to be removed. She was determined to take every chance to retain some sight. The operations caused her great pain and anguish and unfortunately were unsuccessful. The continuing loss of sight was gradual. At one point, she used a television-screen-size magnifier to read. Eventually, she lost sight in both eyes.

Maryellen began to experience kidney failure and was on dialysis for more than a year before she began to explore the possibility of a kidney transplant. Tests showed that her sister Irene's kidney would prove a compatible exchange, and' the operation was performed. It required repeated treatments over several years to forestall a threatened rejection, but finally the transplant became fairly well established and no longer showed signs of failure. In this same period, she had another operation at the Mayo Clinic in Rochester, Minn. Her sister donated a part of her pancreas, but within two weeks it was rejected.

In 1975, I took partial retirement. I needed to take Maryellen to numerous doctors' appointments and hospital visits. During one period, we were seeing from three to seven doctors each week.

During her stay at Mayo, an infection was discovered in the big toe of her right foot—an extremely dangerous condition for a diabetic because of the likelihood of gangrene. It wasn't many months after her return to California that the toe had to be amputated, then her right leg below the knee. She was fitted with an artificial limb, learned to adapt to it, and got around with the aid of a cane. In one year, she had five major operations.

With the loss of her sight, Maryellen lost her ability to

balance, so her movement was restricted to a wheelchair.

Throughout this long ordeal, she was unfailingly courageous. She was very much aware of what I had been through with Althea's illness and did not want to cause such an experience again. Her frequent companion—and a great help to me as well when business interferred with my taking Maryellen to doctors' appointments—was Nancy. But I was with "Sweetheart" through all her major surgery.

Her health began to fail in other ways. She developed angina. Her strength was ebbing, and her weight slipped to 89 from 117 pounds.

Several times I had to rush her to the hospital when she passed out from incorrect doses of insulin. There was no stability to her condition, and it was difficult to establish just how much insulin to administer. I bought an electronic device with which I could measure her blood sugar for the injections I gave her several times a day. Dr. Howard Rosenfeld of Valley Presbyterian Hospital worked with us faithfully throughout this period.

For most of the ten years of our marriage, it was one long disheartening struggle against a siege of failing health. I did not escape untouched, myself. There had been the operation in 1970 to remove half of my stomach—and with it the recurring ulcer problem. Later, there was also the uncomfortable and not uncomplicated necessity to remove a piece of bamboo I'd accidentally driven into my lower colon. And I required a triple-bypass heart operation. In my case, the operations were successful, and my health was restored.

But Maryellen spent the last year of her life almost totally bedridden. Her loyal visitors until the end were Faye Rich, Ben's wife, and Nancy.

Because of what I, myself, have been through with hospitals, doctors, and the medical world, I have determined to make it easier for others who have their loved ones in intensive care, or under treatment for serious operations, incurable illness, or whatever problem that requires continuing and recurring presence in a hospital. I am funding at St. Joseph's Medical

Center in Burbank the building of a hospice with about 20 simply-appointed rooms where family members can sleep comfortably or just rest, bathe, have telephone and other facilities, and be nearby for comfort and reassurance. It should be operating by the mid-'80s.

Knowing that she was dying, Maryellen told me she felt as Althea had that I should not remain alone. "Sweetheart" died in Encino on October 13, 1980.

We buried her in a beautiful setting on the side of a grassy hill overlooking the San Fernando Valley. Many friends and my colleagues from the Skunk Works came to express their sorrow. As they left one by one and in small groups, I found myself standing alone by the gravesite. Nancy noticed this and came up to join me. We walked back down the hill together. She had been a solid support to me as well as Maryellen for the last seven or eight years. She is a beautiful brunette whom I had come to know as an intelligent and admirable person. I realized that I still needed her with me. A scant month after Maryellen's death, I asked her to marry me. Nancy worried that it might seem too sudden. My answer was that while I would be sorry if anyone felt that way, I was not concerned with impressing other people. Life was too short. I had done my mourning for Maryellen through the last years of suffering with her.

"Let's put the past away," was my persuasion. "I don't have time to wait as a mere matter of form. Let's get on with life." Nancy agreed and we were married in November of that year.

18

Defending Ourselves

THE DANGER IN PLANNING OUR NATIONAL DEFENSE is that we prepare to fight World War II all over again. The victor in any future war will have learned that lesson. If there is a third world war it will be a great deal different.

Is the defense currently being programmed for this country really effective? Does it look far enough ahead? Does it risk more than necessary? Are we getting the most for our money? Is it costing too much? Is a "prohibitively expensive" defense really so? Do we want to go down in history as the richest nation in the history of mankind—but be destroyed?

Or is it possible that the realization that no country can afford the defense necessary against new technology might so affect diplomacy that war really does become unthinkable?

The history of the human race does not offer much encouragement. Civilizations have been devastated before with an "ultimate" weapon.

The invention of the longbow and then the crossbow were as important to warfare in their time as the atom bomb or laser and particle weapons today and tomorrow.

When a man fought astride his horse bareback, with only knee pressure and a pull on the mane for control, any peasant could pull him off, stab him, or knock him out with a stone ax. But when the horseman developed a flight control system—a bridle, then saddle, and stirrups—war became darned dangerous for someone on foot.

The invention of the English longbow that could kill a

French knight in armor from a distance of 1,300 feet so shattered the mores of the time that the Pope declared in effect, "Cursed, ye who use the longbow." It was unthinkable that an unworthy peasant could overcome a noble knight. The longbow had a rapid-fire capability, too. In their first use of the longbow on the European continent, the English decimated the French at the Battle of Cressy in 1346, when their marksmen could launch arrows in waves, each shot requiring only a few seconds.

The Turkish crossbow, though slower, had more power—being cranked back mechanically—and could send an arrow slightly farther.

It wasn't until 1803—although the rifle dates from the 15th century—that the Kentucky long rifle, invented by a Pennsylvania Dutchman, could deliver a greater impact with higher accuracy than the English longbow. But the real reason the rifle then became important on the battlefield was not that it was so efficient at killing people but because it made so much smoke and noise that it frightened the horses!

The use of mustard gas in World War I was viewed as so terrible a weapon that all nations agreed to outlaw its use. This restraint was observed in World War II. But then, the gas would have been difficult to control in dispersion and not effective enough militarily for the user to face the inevitable international opobrium. Since Korea, however, use of nerve gas has been reported on more than one side. The morality of man on record does not, I fear, hold out much hope for an end to deadly human conflict.

The technological battles of today will determine the outcome of any future world war. It will be won with new weapons—lasers and charged particle weapons for defense, "stealth" technology to make attacking aircraft invisible, and space satellites for navigation and missile firing. Computer capability may be the most important element of all to winning the conflict, being the controlling technology, insuring the accuracy of weapons firing.

We must not sell our technology. We must not sell, for

example, our best electronic gear—the silicon chips and galium arsenide chips that give computers a memory of millions of bits of information for guidance of missiles, aircraft, submarines, and satellites.

Computer technology is a field in which this country has led for some time. It will be fundamental to our defense against the intercontinental ballistic missile. With enough power, beams can be directed to destroy incoming targets from space bases or from earth bases. These targets—as many as 12 warheads on each missile—must be detected and destroyed with near-100 percent reliability while they still are above the earth's air blanket, well over 100 miles up.

They must not be allowed to get low enough so that the blast to destroy them creates fallout. Even a low-level blast could destroy our own missile bases and cities. A direct hit on earth, with the resultant dispersion of polluted dirt and debris, would be devastating.

Our navigation satellites are fundamental to guiding our submarine-launched missiles with the same accuracy as missiles launched from fixed-ground locations. If we cannot protect our satellites, we cannot insure the accuracy of our missile firings.

In the battle for technology, it is not only what we do but what we do not do that will be important. Our defense can be endangered by actions we fail to take. Failure to develop supplies of critical material. Failure to exploit the resources we have. Failure to think innovatively. Inadequate basic research and development. Insufficient attention to training of technicians, engineers, and physicists. Failure to stop technology transfer.

When we were fighting on the same side with the Russians in World War II, there was a considerable open exchange of technology, of course. They had as good or better equipment than ours in some cases. Our tanks were not comparable to theirs in winter. Their aircraft were better winterized than ours and were operating in freezing weather when we could not even start ours. Winter was a familiar friend for them. On the other hand, our tanks were desertized to operate in Africa, whereas the Russian tanks in the desert would grind to a halt in no time at all.

There also was some inadvertent exchange. We found that when one of our aircraft would have to make a forced landing in Russian territory it would be very difficult to get it back. We tried hard to avoid that. They did copy from two B-29s forced down and retained there.

The United States has been slow to tighten security on access by Russia, especially to some seemingly simple but strategically important basic technology.

Concrete hardness testers, for example, would not seem at first thought to be strategically important. They are used in this country to determine strength in a bridge or roadway—or a missile installation. The tester tells us what kind of weapon it would take to knock out an installation. Several were sold to Russia before supply was cut off.

Gear-shaping equipment that has made our submarines more quiet for years than the Russians' has been sold to them. So has ball-bearing grinding equipment that could improve their missile-firing accuracy by a factor of eight or ten.

Our own Air Force some years ago received an award for size of a load carried in a single airplane—40,000 pounds of switching gear flown to Russia in a C-5 cargo plane. And how would that gear be used? It was capable of switching tremendous amounts of power in nanoseconds, a necessity in magnetohydrodynamics—generating high-powered rays electrically or from nuclear sources. The Russians needed the American equipment to generate the very short time pulse that is the basis of what we believe to be one of their new weapon systems.

It was no secret that they were undertaking four to five times more work than this country in the field of lasers and charged particles—commonly called "death rays"—the next major weapons. Should the Russians develop the capability first to make our missiles impotent, there won't be a war, just a surrender. They may be ahead of us in charged particles. I think we may be ahead in lasers. I'm quite sure we are ahead in infrared use. But I do not think we should make it any easier for them by transferring technology in any of these areas.

There was interest by the Russians during the early '70s in buying Lockheed's L-1011 transport. It was the latest in advanced passenger airliners. The Russians wanted to buy three planes only. This would have provided them with three complete sets of drawings and all manuals, including details on the world's only advanced automatic blind landing system. That would have been very useful for all-weather bombers. It would have been a very inexpensive way to acquire the technology without a long research and development program. And the Rolls-Royce engine on the airplane is much better than anything the Russians have. I was one who protested that sale, and I must not have been alone. Somewhere along the line the deal was dropped. The English are reported to be considering sale of the engine still.

Future military aircraft will be very expensive. An entire new fleet—fighters, bombers, ground-attack airplanes, cargo carriers—designed with the latest "stealth" radar avoidance techniques—would cost more than this country could afford realistically.

We will have fewer types of advanced new models and fewer of them in number. Because of the high cost, it becomes critical to deploy them only on key missions. Vulnerability of the new systems can be lessened and effectiveness increased by mixing them in service with the large number of old and obsolete models—manned on support missions, or unmanned as bombers, missile carriers, or drone decoys.

In new design, we must not look backward and try to put maneuverability in the airplane over all else, but rather put it in the missile. We may not need to endanger a man in the vehicle at all on the most hazardous missions.

It will be a very selective process, deciding how to fight a future war. Superior performance will be required of the new systems. And toward that end, work needs to be done in several basic fields. It should not be forgotten that the major aeronautical advances of World War II were not ours but German—the swept wing, the delta wing, and the jet engine, for example.

Work needs to be done to give our fighter aircraft more range in supersonic flight than they now have flying subsonically, and without afterburner and its extravagant use of fuel. The range of the F-15 in supersonic flight at sea level is about 57 miles. The F-14 doesn't fly much farther. The extra range will come with improvement to the type of engine built by Bristol and used now in the Concorde transport. This engine shifts cycles, using a slight amount of afterburner to boost speed to supersonic, then cuts off the afterburner to cruise at Mach 2 with very economical fuel consumption.

Another area requiring more research in the transonic range—speeds from Mach .9 to Mach 1.1. In this speed range today, aerodynamic drag goes up by a factor of 300 to 1,000 percent, with tremendous compressibility effects. We still rely on primitive forms of dealing with this phenomenon. We have learned how to deflect it, but not yet to conquer it.

There are ways of minimizing it. In the F-104, we did it with a razor-thin wing. It can be accomplished also with highly swept-back wings. In the YF-12A, the brute power of the engines just pushes the plane through the transonic range. But these are not efficient methods of solving the basic problem.

Fundamental research is needed. The logical agency for this is NASA with its excellent research facilities, representing the investment of hundreds of millions of dollars, and which it would be a waste for private industry to duplicate even if funds were available.

It is a waste of our resources, too, when research is repeated. Yet this occurs. Two specific examples: development contracts for aircraft radomes able to withstand 500°F temperatures; a contract for development of titanium landing gear. The Blackbirds have been operating with titanium gear for 22 years! Their radomes give fine performance at 650°F!

There also is not enough use of what we already have accomplished. Sometimes the attraction for something new is irresistible over adapting proven equipment for a lot less money. We should not be repeating costly development work. Lockheed's Lancer and universal trainer proposals, discussed earlier, come to mind. Rather than improve the proven for readily-available, low-cost vehicles, the military opted for new aircraft with comparable capability to be developed over a considerably longer period of time and at much greater cost.

We must study our areas of potential vulnerability. Are we relying for defense on a team of dinosaurs? If it is necessary to penetrate an enemy country, what will be the best way to do it?

Many mappings from U-2 overflights and space satellites provide us with information on the location of Russian radar and missile stations, sites of factories, and other strategic targets, for example.

A number of years ago the Skunk Works made studies of penetration into the Soviet heartland. We computed probable aircraft losses for different approaches—bombers coming in at sea level to others operating at 80,000 feet. The study did not incorporate aircraft with low radar cross-section design.

We evaluated the proposed B-1 subsonically at low level and a hypothetical bomber cruising at Mach 3 at 80,000 feet.

Our conclusion was that a subsonic plane at low altitude would be subject to attack by all versions of Russian fighters, from the older MiG-15 to the later faster types. Efforts to incorporate the latest radar avoidance techniques in already existing design were not very productive. The loss rate was put at 35 percent of the fleet. And cost per unit for the bomber was more than $200 million per plane, not counting costs for crew training and support.

The high-altitude supersonic bomber was much more expensive than the low-altitude aircraft, but had a survivability rate about three times greater.

My own conclusion from these early studies was: why a manned bomber at all? If we can get the accuracy we expect from intercontinental missiles, I see little reason for sending a man on the attack mission.

A familiar argument is that the bomber can be recalled. Well, the missile can be blown up en route with a radio signal.

There is little reason for putting a man over Russia except perhaps for reconnaissance in some cases. And then you hope he will survive the 45 minutes' overflight through high-altitude clouds of nuclear contamination.

One U-2 monitoring atmospheric quality a few years ago, when both Russia and the United States were testing hydrogen bombs, found the same cloud of nuclear debris circling the earth six times—propelled by the jetstream along an airplane polar route over the United States.

The vulnerability of the U.S. Navy, or any navy, in a nuclear war—or any kind of war—is a concern. With satellite tracking stations making a pass overhead every 90 minutes, it is very easy to follow a fleet moving at a speed of 20 knots. At one time, Russian satellites actually were providing us with much of the information on our fleet location. Our own reconnaissance was better than theirs over land.

It is perfectly feasible to launch a land-based ICBM or IRBM carrying a dozen warheads at a fleet under way. I know of no way at present to stop an incoming missile speeding on a 90 degree course straight down on you.

The Russian Backfire could launch its missiles, carried under the fuselage, from about 240 to 250 miles away and guide them to knock out our capital ships. The ability to stop that Backfire is important.

The vulnerability of the U.S. Navy is vitally important. Military missions aside, a high-priority purpose is to protect the tankers hauling oil around the continent of Africa and transocean from the Middle East. Keeping these shipping lanes open is important for more than one reason. There is in this country a shortage of strategic materials—e.g., vanadium, chromium, platinum. Many of these metals come from Africa and the developing nations.

Russia has very good submarines—bigger, faster, deeper diving than ours—and many more of them than we have. Their newest is almost as large as a cruiser. Their subs can do 50 mph—much faster than ours. Their latest have titanium hulls which make them more difficult to detect. Titanium is nonmagnetic and escapes some of the means we have of detecting submerged submarines. The Russians can build these large-size titanium hulls that give their subs the deep diving capability because they have the huge presses to do it. We do not.

If Russia with all her submarines decides to put us out of the shipping business, it will be a big problem for us.

Of course, we may find other ways to match their submarine performance. Do not write off our Trident and earlier Polaris submarine-launched missiles.

Anti-submarine warfare is a constantly changing battle. Lockheed's carrier-based S-3A ASW aircraft for the U.S. Navy, after only five or six years' service, already had changed over to new electronic gear for locating submarines.

Several years ago it was discovered that every submarine makes its own distinguishing operating sounds. No submarine is totally quiet, though that is the goal. These sounds now have been classified into a sort of directory, so that with sonar and other detection equipment our ASW planes, ships, and land-based stations can follow and identify the trail of an individual sub, know whether it is large or small, diesel-powered, electric, or nuclear.

ASW aircraft really originated with the Hudson bomber during World War II when an RAF plane became the first in history to capture a sub. Lockheed since then has built more ASW aircraft than all other companies combined.

ASW has been a developing science from those first beginnings, when the target had to surface with snorkels to recharge batteries to today's nuclear-powered models that can remain submerged for days. Historically, the submarine has been ahead of the game step by step, temporarily to be overtaken by search-and-destroy techniques but then racing ahead again.

For years, I have said—jokingly because it is totally impractical—that in any next war I wanted to be in a nickel-plated, nuclear-powered, deep-diving submarine with plenty of food and reading material, because it would be the safest place in the world. Nickel plate would make the sub very smooth and very quiet. That would be prohibitively expensive, of course, but there are other platings we are studying seriously now for silencing purposes.

"Operations analysis," or "operations research," as an approach to design decisions really took off from those early ASW efforts in World War II and immediately afterward. Lockheed determined to stay in the anti-submarine business, and to do so we knew we had to keep ourselves educated. After the war, we sought a Navy research contract, even on a "no-cost" basis. I set up a group under Robert A. Bailey to study all phases of submarine and anti-submarine development—sonar, weight analysis, noise, etc.

We were given privileged information and, in return, reported to the Navy every few months on results of our studies. The heart of operations analysis, and the only method to make it worthwhile and accurate, is to keep it a purely research effort. Never use it as a sales tool. In the long run that is counterproductive, because it leads to tainted conclusions.

There are other ways our enemies could interrupt our vital supplies—such as subverting the governments in the developing countries that supply much of our strategic imports and installing governments sympathetic if not subservient to their own.

Development of sources of basic materials where possible in this country and others in this hemisphere is especially indicated because of these threats to our supply.

The titanium "sponge" from which the sheet and bar were formed for the SR-71s came principally from Australia and Japan which have it in good supply. But the basic materials for the later Blackbirds came also from Russia, which had developed its titanium-producing facilities and decided to undercut the others in price. We discontinued those purchases, however, after an initial one because we did not want to help Russia develop this industry.

The titanium ore found naturally in this country is not the rutile from which the basic sponge has been made to date. It is a different form of titanium oxide—ilmenite. It has been cheaper to buy from foreign sources in the past rather than develop the local product. While it will require more power to process our native ore, it should pay overall to insure availability of the metal. We know how to do it, but the expense of the necessary investment has delayed development so long because importing the product was cheaper.

This comes back to one of my favorite crusades—developing a titanium capability in this country and getting the cost of the metal down to where it is reasonable compared to other materials. This means mining and processing the ore, building rolling mills and sheet metal plants, and, especially, building a big enough press to forge the large submarine plates that give the Russian subs their deep diving capability, and other large production pieces, such as aircraft landing gear.

The initial cost would be tremendous for such a press alone, but the value in availability of the material, time saved in production, quality of the finished product, as well as importance to national defense must be considered to overbalance the dollars involved.

One of the most important things we can do in the battle for technology is to train young engineers, scientists, and technicians who can follow the tremendously complicated and complex new programs. We are short of technicians, especially. And for dealing with the technology of the future, we cannot quickly reassign engineers from conventional aircraft design. Engineers still will be required to design and build our defense systems. But the discipline now that will determine what these are is physics.

The Russians are graduating five times as many engineers each year as the United States. There is no unemployment of them. Here, unfortunately, there is little or no stability in our programs. It's train, hire, and fire.

The defense of this country and the Free World requires an operations analysis approach—looking at the entire area from scratch, objectively. What would a war be like? Nuclear? Nonnuclear? What weapons will we really need? Expensive nuclear-powered aircraft carriers which might last two or three days? Should we put the carrier underseas—as a submarine?

Do we need manned aircraft when a missile can be fired and controlled accurately from the ground? Should we use our old obsolete aircraft as decoys while the new highly-sophisticated and very expensive technologically-advanced models head for target? In the operations analysis approach, no idea is too outlandish to consider—and then evaluate for effectiveness, cost, complexity, flexibility, reliability, manageability, and all the other characteristics that come into play.

In the Skunk Works we have a dozen or so people working at all times in this manner, keeping ourselves educated on what "they" are doing and "we" can do. How good are their surface-to-air missiles? Their radar? Their next airplanes? Their research and development in other fields? How do we penetrate the country in case of war?

This approach as a national policy is basic to defending ourselves.

19

Technology and Tomorrow

By the year 2000, the "death rays" of the comic strips and science fiction will be a reality. Laser beams and charged-particle weapons will be our defense against enemy missile and rocket attack in any nuclear war. Computer-controlled, they will detonate the incoming warheads in space.

That is the scenario as written today. Accomplishing this will be no small task.

Lasers travel at the speed of light, more than 186,000 miles per second. While there are peaceful uses of the laser—in surgery, manufacturing, and other industrial applications—in lethal weaponry the laser gun will be able to pick off incoming rockets traveling at speeds anticipated to be as high as Mach 24.

First use of the laser in defense is envisioned as space-based, because laser beams in practically any frequency fall off greatly in effectiveness as the atmosphere deepens.

While enormous power will be needed to place and operate a laser weapon in space, it will require even more to fire it from the ground through the atmosphere.

Our first defensive weapon against nuclear missile attack, therefore, should be a very sophisticated ground-launched anti-ballistic missile. This I believe we should develop as soon as possible despite Salt I or Salt II. And concurrently we should develop what I believe will be the most effective defense against incoming enemy missiles—laser or particle-beam weapons located in space.

Not only is this possible, it is necessary.

Our initial efforts at finding a target and aiming a laser gun are very clumsy, but we have succeeded in hitting a target from a C-141 flying at low altitude. The system once developed probably will use a combination of infrared, radar, and electro-optical systems.

Operated from a string of perhaps two dozen satellites in orbit, lasers would provide the capability to detonate from a few to several hundred rockets still in their launch and boost phases. The defense weapon must not only be fast to intercept the incoming charge but able to do it repeatedly and accurately, switching from one to another of a large number of targets fired at once. This will require the world's best guidance system.

Charged particles are a form of nuclear weaponry, but without contamination. Practically no mass is released, just energy. It employs the science of magnetohydrodynamics—the flow of high-powered rays generated electrically or from nuclear sources. Very high speed electrical energy can be developed with electrons beamed and released from an electrical container. The method of entrapment and release of the charged particles requires temperatures equivalent to that of the electronic activity on the sun—hundreds of millions of degrees centigrade. Generated here on earth. But the period of time involved is so short—a fearful jolt in a nanosecond—that total power is negligible.

We do not know yet what these weapons will look like. Essentially, they will be huge generators that will form an electronic containment. Various gases would be injected to develop electron flow. Releasing this flow is a very difficult development. And it is in this field that the Russians are using that switching gear—capable of transferring tremendous amounts of power—that they obtained from the United States in one of the technology exports that I deplore.

I've always liked to think of this force, lasers or particle weapons, as creating an enormous teepee around our own targets—missile sites, large cities, government seats, for example—a teepee rising from earth above the atmosphere so that nothing can fly through it. Any nuclear bomb aimed toward us would be detonated in space without the resultant fallout in the atmosphere. It takes the atom to defeat the atom. We would need to generate a tremendous amount of power, of course, to erect such a ground-based protection. We are working on it.

The role of satellites operating in space will be vital. Especially important are the navigation satellites fundamental to our guidance of submarine-launched missiles, the Polaris and Trident. They will provide the same accuracy for submarine-launched missiles as for firings from a fixed point on earth. Within just a few years, well before the year 2000, we expect to be able to fix a position any place in the world within ten feet. Both lasers and particle weapons will be a necessary defense of these satellites.

What will be the importance of aircraft in the year 2000? For defense? For commerce?

It may seem traitorous from an aircraft designer, but I see a diminished role for the manned military aircraft and more reliance on remotely piloted vehicles and missiles. When you can put 20Gs of maneuverability in a missile while a man can pull only 9Gs at most—nine times the gravity of his own weight; when you can provide a missile with the search capability to find its target; and when television and other relay links from a high-flying U-2 or space satellite can give rapid readout in real time to a man at controls in Washington, why send a man over enemy territory at all?

If we do use manned fighters and bombers against ground targets they darned well better be invisible at any flight altitude because of vulnerability to ground-to-air missiles as well as other fighter aircraft.

"Stealth" is the technology that will change the character of aerial warfare. If the enemy can't see the aircraft with radar he can't hit it. The capability of fighters and bombers will be enhanced greatly when the flight crews do not have to worry about ground forces except possibly to destroy them.

"Stealth" technology still is being invented daily. We developed and introduced it on the first Blackbird, and the actual shape of that series of aircraft is fundamental to their reduced radar reflection. Also, 20 percent of the surface of the aircraft is made of "stealthy" material. But those planes still rely on other elements of design to avoid detection—altitude, speed, and electronic jamming capability.

The technology no longer is entirely Lockheed proprietary. This industry is no worse nor better than others in competition for business. The number of unsolicited proposals for which the Skunk Works has been awarded sole-source contracts has aroused a good deal of envy. I'm proud of our record, we've earned it. And government policy is to maintain competition among aerospace companies. That is understandable, because it keeps us on our toes to try to stay ahead of the others.

So, despite the strict security imposed by the Skunk Works and the military on its employees, recent retirees from our company and certain key agencies now are enjoying exceptional job opportunities elsewhere in the industry; sometimes with as much as 60 percent more salary, stock options, an automobile, or other bonus, for part-time work as consultants. What do these high-paid retirees have in common? They all worked on "Stealth" technology.

The year 2000 is less than 20 years away, and I personally do not try to project much farther ahead than that. Who would have predicted in 1938, for example, that we would be flying three times the speed of sound by 1958? Of course, this was developed in great secrecy and there still were those at the time who said it couldn't be done. Or who would have anticipated that in only five years, from 1977 to 1982, the cost of a jet transport would rise 300 percent? Or that jet fuel would sky-rocket from seventeen cents to $1.50 a gallon?

In retrospect, this country was wise not to have gone ahead with its supersonic transport in the 1960s. And Lockheed was fortunate to have lost that design competition. The SST would have hit the fuel crisis head on. And the noise would have been unacceptable. It is not an airplane we can afford to fly today in commercial use. The Concorde, of course, enjoys government subsidies by Great Britain and France.

The Lockheed SST proposal basically was a three-times

scale of the SR-71 design, already proven in flight. We did not use the wide fuselage so beloved by the airlines for the variety of interior arrangements it can accommodate because of what we knew about the importance of weight and drag to attaining triple-sonic speeds.

The airlines opted for the wider fuselage of the Boeing design. And the contract, later cancelled, went to a company that had never fired an afterburner nor made a sonic boom—that is, had never had any supersonic experience. We made our design studies available, and the Boeing plane came to resemble more and more the losing Lockheed proposal. But their design at the time of contract cancellation lacked transatlantic range by 700 miles. I wanted the concession to pick up the passengers from mid-ocean.

Very high fuel consumption still is a problem for commercial operation. To be economically sound, an SST will require development of another series, or two, of jet engines with much greater thrust-to-weight ratio that can achieve supersonic speed without afterburner. Whether we develop these improved engines by the year 2000 is dependent on availability of development funds.

And for successful commercial airline use, the supersonic transport first must overcome the noise problem. This, too, will yield to advanced engine development.

There is a technique, not a solution, that could be used right now to reduce takeoff noise but I have not been able to persuade others that it would be acceptable to passengers. The passengers wouldn't even know when it was taking place. I refer to mid-air refueling. We had done it with the Blackbirds more than 18,000 times by the early 1980s.

Sitting in the second seat of a YF-12 in flight, I have been amazed at the speed and skill of refueling from a KC-135 tanker, even when the aircraft made turns, climbs, or other maneuvers. This mid-air refueling process I believe to be one of the most important developments in the history of aviation. Why? Without it we could not send our bombers over Russia—and back, for example. We could not send great payloads over-ocean as we now do with the C-5. Nor could we send our fighters halfway around the world. Nor cross 7,500 miles of the Pacific Ocean in five hours with the SR-71. And the mid-air refueling capability eliminates the need for many foreign air bases.

Using the technique, an SST could take off lightly loaded with fuel, take on enough fuel in the air for the flight, and land conventionally. The present Concorde, for example, could take off from Los Angeles with a full load of passengers but light in weight and therefore not requiring the noisy afterburner, refuel over Hudson's Bay from a 707 or other obsolete transport used as an aerial tanker, and fly nonstop to London. Realistically, I do not expect that to happen.

But this is an area where I do not expect the Russians to surpass us. There was an amusing incident involving the Russian SST at the Paris Air Show in 1973—before its tragic crash. Visitors were invited to inspect the airplane on the ground and in flight. Other Lockheed people were allowed to go aboard and even to fly in the airplane. I was escorted around the outside by eight of their engineers. They didn't seem to hear when I said that I'd like to go inside. So I had a good view of the exterior.

While I have been impressed with the Russians' forging capacities with their big presses, presses we need and do not have, I found that manufacturing techniques applied to the airplane skin were quite crude. Rivets were not flush with the surface. The fuselage was as well made as it needed to be, but the infinite pains routinely taken by U.S. manufacturers to get the skin smooth had not been taken.

We have been surprised to learn of the lack of concern for safety in some Russian designs. Their standards—certainly in military aircraft and even commercial airplanes—do not meet those of the U.S. Many of their airliners could not pass an FAA test for engine-out takeoff.

Military aircraft in the Korean conflict did not have an abort speed. A four-engine bomber would be so loaded as to require the entire runway length for takeoff. If an engine were lost there would not be enough extra runway for takeoff on only three engines. It is not that the Russians could not improve in this area, it is that they have chosen not to do so and have their priorities elsewhere.

The HST, hypersonic transport, would be the next step technologically. Practically, it really is difficult to make a case for going to Mach 4 and 7 in a transport airplane. It takes too long to get up to speed and then throttle back for landing. Less than one-third of the flight duration would be spent hypersonically.

The HST would seem to have no place at all in the commercial airline field. Even on very long flights—the only flights on which this airplane would be used—our best knowledge today indicates that about 37 percent of the flight distance would be spent in climbing to altitude and accelerating to design speed, where the plane would cruise for about 30 percent of the flight distance before starting its descent.

Fuel consumption for the hypersonic engine, as we know it today and anticipate it for tomorrow, really is out of this world.

As a passenger airliner, the HST would not be economically feasible. In military applications, it probably would be unmanned. And the SR-71 already exceeds Mach 3 and altitudes above 100,000 feet.

The nuclear-powered airplane is another concept considered for the future. Under contract to the Strategic Air Command shortly after World War II, we investigated design of a nuclear-powered bomber. It was even before our first Skunk Works was established. I was chief engineer then for Lockheed's California Company.

Gen. Curtis LeMay, then SAC commander, wanted a plane to fly high and supersonically. It was the NEPA project, nuclear energy for propulsion of aircraft. Six or seven other companies were involved. James Douglas was Secretary of the Air Force at the time, and 30 years later when I saw him in Washington, he came over and thanked me for "cutting up" the nuclear airplane.

We had been asking nuclear power to do something it was

not suited for, and I said so. The airplane design became mammoth in size to carry the big nuclear powerplant. The cockpit alone weighed 40,000 pounds. A lead shield was required between the cockpit and rear of the airplane where the reactor was carried in order to reduce the radiation enough to allow pilot and flight crew to fly the plane for about 30 hours a year.

It was so "hot" from a radiation point of view that if you had to change a generator on one of the engines—either four or eight in the design study—it had to be by remote control, by a robot. The plane really didn't like to get off the ground, so jet fuel afterburners were necessary. It got to be a great big cumbersome unwieldy system.

Funds had been approved to continue the work, but I argued against it. After some strong discussion, others reluctantly came to agree with me that the project should be abandoned.

I do not foresee the nuclear airplane in the year 2000 either.

The Space Shuttle is an exciting concept with lots of popular appeal. Philosophically, putting a man in space and bringing him back down means a great deal to our belief in ourselves. Whether the Shuttle will pay off economically, commercially, I do not know. The number of flights per year and realistic pricing for customers are yet to be determined.

Its continued safety concerns me on these early flights. It was not all easy going on the second flight when the crew was down to the last of three power sources for electrical needs.

We can do so many things with an unmanned cheap launching device. Communications satellites have been orbiting for several years and do a marvelous job. They are a very good business proposition.

On a more mundane level, except for development of second-generation jet transports, commercial passenger, and cargo aircraft in service today—or planes much like them—will be what we have in the year 2000. The emphasis will not be on bigger and bigger as it was for some years, but on what is realistic, practical, and commercially viable using present technology.

One of my favorite ideas for a number of years has been a method of sinking capital ships without using a nuclear or even a gunpowder bomb. It would be truly a "clean bomb." I thought of it when the *Pueblo* was captured, and the *Mayaguez.* We could have sunk the things without hurting a soul once our people were off.

If a 2,500-pound highly streamlined shape made of tool steel—which would not shatter—were to be launched from altitude by an SR-71 it would hit sea level at speeds well above Mach 3. Its penetration power would take it right through any ship, and it would generate so much heat that it would set fire to or sink the ship. It would be a clean kill, and much cheaper than a conventional weapon.

Such a bomb could go through 300 feet of earth. It could, for example, plug the tunnels through the Ural mountains. It could penetrate 33 feet of reinforced concrete.

We know the airplane could carry and launch such a bomb because of our earlier missile-firing work with the YF-12.

The bomb must be made of tool steel to be very hard and not break apart on impact. The trick is in the guidance, and we would expect the bomb to hit well within a thirty-foot target area when dropped from 85,000 feet. It is quite easy to figure the penetrating force—with the weight and drag and the force of velocity. Design of the weapon itself is quite simple. And I'm not giving away an idea to the Russians, because they haven't anything that can fly high enough or fast enough to launch it.

There is promise of resource development beneath the oceans with new techniques in ocean mining. We know the resources are there including the chromium, vanadium, platinum, and other scarce materials that now must be imported from Africa. The techniques are known, too. It would require only the investment of funds, hundreds of millions of dollars, to make it a practicality.

The *Glomar Explorer,* a project of Hughes Aircraft, Global Marine, and Lockheed Missiles and Space Company, was designed for ocean mining. A sort of vacuum sweeper reaches down two or three miles to pick up nodules from the ocean floor. This has been done in test operations under ocean west and southwest of Hawaii. A processing plant aboard ship can reduce the nodules to ore. The technology is there, although to be settled, of course, is the ongoing argument about which nations reap or share the benefit of products from international waters.

The *Glomar* has another very important capability—rescue and retrieval of submarines.

Lockheed's participation in *Glomar* was to design the mechanism that would pick up an abandoned Russian submarine sunk to depths of 15,000 feet. Skimping on static testing of the remotely-controlled titanium arms—failure to conduct one last test before the retrieval attempt—resulted in less than 100 percent success. The sunken submarine had been located and was being lifted. It was two-thirds of the way up when one of the arms failed, and part of the sub dropped back down. The rest was recovered, however, and it was informative to our submariners.

Later, one of our submarines was lost in the east Atlantic. We suspected that it might have been the victim of the game of "chicken" the Russians like to play with other subs. But there was nothing capable of descending to 9,000 feet to search for it—even for inspection if not retrieval. So the value of the vehicle is indisputable—militarily and commercially. The expense of restoring *Glomar* would be great. Just the cost of maintaining it in dock runs high—about $30,000 a month. But its usefulness is clear. It would be a handy gadget to have in operation.

Not all of our weapons are military. Some are economic. The most important airplane for the future, to my way of thinking, isn't a transport, isn't a bomber, isn't a fighter. It is the crop duster. Why? We are going to have to feed an awful lot of people in this world. We must keep our ecology in hand, save our forests, seed the fields, fight fires, control weather, and even—should there be nuclear explosions and environmental contamination—spray to accelerate diminution of radiation.

There is nothing dramatic about this airplane. It just might be the airplane most important to more people than any other. I'd like to think that airplane was one for peaceful purposes.

20

A Good Life

At Star Lane one Saturday morning in a recent spring, we branded 52 calves, ten more than at roundup the previous year. The day began early for Nancy and me as we greeted 64 friends, neighbors, ranchers, and cowhands arriving to work or watch and share in the traditional barbecue that followed. It is one of the regular rituals we enjoy on the ranch—probably our favorite.

The roundup actually began the day before. My neighbor rancher Dee McVeigh brought nine of his cowboys to join our hands in combing the hills to round up all the cattle. All but one wild cow, that is, which escapes us every year. She's more deer than cow, she's so fast. We go after them all—calves, cows, and steers.

Some of the steers and the dry, non-bearing cows will be sold. The new calves, six to eight months old, will be sprayed for flies and ticks, innoculated in one mixed shot against hoof-and-mouth and several other diseases, have their eyes sprayed for pink eye, and be castrated and branded—a star with an L offside. All the cattle normally get shots once a year and are sprayed twice.

Before day's end on Friday, all the cattle had been driven into several holding pens awaiting the next morning's action.

On Saturday by 8 A.M., we were organizing the hands into roping groups—several teams of three or four men on horse-back. Their first chore was to separate the calves from the cows, then divide the calves into groups of about ten. One group of calves at a time was herded into the main corral for the job at hand. We change cowhand teams with each group of calves.

Halfway through the morning we take a break to have coffee, soft drinks, and sweet rolls.

This is hard work. Yet many of the men are gray-haired and over 70 years old, a few 75. They've been roping all their lives. As in piloting, experience is what counts. There are some young ones coming along, too, among them our foreman's son, Larry Erickson.

The branding irons I handle myself. My good friend of many years, Dr. Lowell Ford, handles the surgical assignment, aided by a young woman who is a veterinarian's assistant. Dr. Ford, no veterinarian, joins us every year just for the fun of participating in the roundup. He comes from Kernville now, where he has helped to establish a much-needed clinic. A versatile man, a humanitarian and intellectual, he has taught philosophy of religion at Occidental College. He found this totally compatible with the practice of medicine.

By the time all the calves have been worked through, the job is done for another year. It is mid-day, and time for the barbecue.

As a working hand, I retreat to the main house to shower and change and return in my "padrone" outfit—western pants, embroidered shirt, and big Mexican sombrero. That is my role for the afternoon.

The barbecue setting is a grassy hollow under the shade of several huge ancient oak trees. The barbecue pit is a permanent fixture here, and the coals already are glowing. Long tables with benches have been set up. A bar has been created from a large box set on end; and soft drinks, beer, wine, and stronger libations are dispensed. An array of dips and chips is spread over one of the tables.

It is a congenial crowd, and newcomers do not remain strangers for long.

After the snacks are demolished and several rounds of drinks downed, the serious food arrives—barbecued sausages, tritip steak, beans, several kinds of salads, and an assortment of desserts. There is food enough to fill any hungry cowhand.

Then we play poker—some of the old hands and I. It's a wild game, with each dealer calling what type of game will be played with that hand. Nobody wins or loses a lot, but the game is a major event on the day's schedule each year.

Before sunset, the group gradually drifts away and we drive back up the hill to our house—Nancy and I and a few family members and friends who will stay overnight. Each time of day and each season presents its own special type of beauty in these mountains, but I think sunset at the end of a satisfying day's work is my favorite time of all.

Even with a full-time foreman, there is plenty for an interested owner to do on a working ranch. Nancy took to ranch life enthusiastically from the first, as had Althea. Both of us are involved with everything that goes on there—reviewing the work schedule with Lee, tilling the soil, planting our oat hay, crop dusting against the invading mustard, harvesting, baling, and storing the crop.

We had a good crop that year, and by late summer had 10,000 bales of hay stored in three barns awaiting sale at higher prices during the winter. Many ranchers have to sell on harvest, because they haven't sturdy barns with cement floors to keep out gophers.

Before all the bales were picked up, however, the whole crop was endangered by a fire that swept over 135 acres of the harvested area. It was started by a spark from the machine that picks up the baled hay—despite a spark arrester, practically new, and twice inspected. Lee had been working late on an extraordinarily hot day in mid-summer. It was almost dark when he shut down the machine and left the field. By the time he arrived at his house, he turned to see flames against the darkened sky.

The interdependent feeling found among ranchers and farmers still, the same spirit of frontiersmen who knew they required each others' help, sent a neighbor to the south of us to drive over with a truck bearing a water tank. He is a man we had yet to meet. The help was much appreciated but not needed.

Firefighters with three helicopters and two water bombers arrived within minutes of being called and put out the fire within an hour.

There have been other fires in the area from this same source. The spark arresters on these machines are made of ordinary galvanized wire screening and burn out in about a month's time. They meet the legal requirements, but obviously that is not good enough. I am designing new spark arresters of stainless steel for our equipment.

Another crop we harvest each year at the ranch is walnuts from about 40 trees. This is not a paying crop, though. It's a Christmas present for friends.

It is out of the question to try to raise a vegetable garden, much as we should like to do so. There are too many gophers that we've never found a way to eliminate.

Our riding horses we take care of ourselves. I presented Nancy with a new Palomino, and she rides frequently. My own horse is ridden so seldom anymore that I have about given up that activity. I do not wish to be thrown and get a broken back at this time in my life.

I do spend a great deal of time rebuilding the machinery—keeping the trucks, tractors, and other equipment in condition. I still love working with machinery. I can spend as much time as I have in my shop. It is so well-equipped that we very seldom have to call on outside help. This has been a big part of my enthusiasm for the ranch—keeping all the equipment running. I think that without it, the expense of running the ranch would be about twice what it is.

Most of the major projects I've wanted to build on the ranch have been completed now. One of the first projects I undertook when I bought Star Lane was to build a combined dam and bridge over the creek that runs through the ranch. I had to decide whether to stress it for just a hay truck or a heavier load such as a Minuteman missile. The missile won out. Anything of that weight could safely be carried over the creeks on my ranch. Should the day ever come when weapons need to be dispersed from nearby Vandenberg Air Force Base, or elsewhere, Star Lane won't be a bottleneck to their being deployed in the surrounding mountains.

My work always has been exciting to me and still is. Very serious study, while demanding, always has been a joy. I literally love aerodynamics, mathematics, physics, machinery—all the tools of my trade. I consider myself very fortunate to have lived my professional life doing exactly what I always wanted to do.

The fantasizing I did as a boy, imagining myself to be Tom Swift, I still do—but now as Jules Verne, stranded somewhere on an island with no ready-made means of escape. As I swim across the pool at the ranch or at home in Encino, I pose problems to myself. How would I make an airplane from absolute scratch? How would I find the ore, dig it up, smelt it? How make steel? How devise an engine, carburetor, ignition system? What could I build? Tractor? Ship? Or, not so isolated as Verne, what could I accomplish from scratch using only the tools at the ranch. The mental challenge is entertainment for me.

Not all of my heroes have been in fantasy, though. There are a few from real life. Dr. Charles Kettering, for many years in charge of research for General Motors, is one. I learned a great deal in just reading about him and how he operated, though I never met the man. He headed an excellent laboratory and made many important contributions—among them the electric starter for automobiles that made it easier for women to drive.

It took a hard crank to start the engine before that.

Thomas Edison is another of the men I most admire. He was such a prodigious inventor and so tenacious in pursuing a goal. He was not afraid to go off into untried areas, not afraid of criticism for doing so. He not only was interested in invention—e.g., the electric light bulb—but followed it to practicality.

My own life has come full circle. It's a long way from the harsh climate of Northern Michigan to the luxuriant mountains and valleys of Southern California. But the same elements are there that I so loved in my boyhood. The ranch is my hideaway in the woods, the horses and dogs are my pets, the shop is my most enjoyed hobby, and, of course, my library is extensive. The same good things—just a lot fancier and more of them.

Once again I have a happy home life. Nancy and I divide our time between Encino and Star Lane. I continue to work several days each week at Lockheed and travel frequently to Washington for consultation on aerospace matters. Nancy accompanies me, and we enjoy the social exchange with many different people. Whenever we have time back in California, we spend a three-day weekend at the ranch.

It's a quiet life there—usually golf at Alisal and dinner out on Friday. Nancy is a good cook and we enjoy the time to ourselves without help in the house. We have our live-in housekeeper, Carmen Loayza, in Encino and our two police dogs, Wolf and Prince. In the first year of our married life we spent a family Thanksgiving at the ranch and Christmas in Encino.

On the way to completing this book, I was detoured by Dr. Jerome Sacks for a triple bypass heart operation—my second. It had been nine years since the first.

The final chapter of my life is not yet written. But if God should call me tonight, I will have had more than my share of it all—poverty and wealth, struggle and success, obscurity and recognition, sickness and strength, sorrow and joy, happiness and love.

More than my share.

Appendix

Awards and Honors

1937   Lawrence Sperry Award, presented by the Institute of Aeronautical Sciences (now the American Institute of Aeronautics and Astronautics) for "important improvements of aeronautical design of high speed commercial aircraft"—for development of Fowler flap on Model 14. Presented annually "for outstanding achievement in aeronautics by young men."

1941   The Wright Brothers Medal, presented by the Society of Automotive Engineers for work on control problems of four-engine airplanes.

1956   The Sylvanus Albert Reed Award, presented by the Institute of Aeronautical Sciences, for "design and rapid development of high-performance subsonic and supersonic aircraft."

1959   Co-recipient of the Collier Trophy as designer of the airframe of the F-104 Starfighter, sharing the honor with General Electric (engine) and U.S. Air Force (flight records). The F-104 was designated the previous year's "greatest achievement in aviation in America."

1960   The General Hap Arnold Gold Medal, presented by the Veterans of Foreign Wars for design of the U-2 high-altitude research plane.

1963   The Theodore von Kármán Award, presented by the Air Force Association for designing and directing development of the U-2, "thus providing the Free World with one of its most valuable instruments in the defense of freedom."

1964   The Medal of Freedom, presented by President Lyndon B. Johnson in ceremonies at the White House. The highest civil honor the President can bestow, this recognizes "significant contributions to the quality of American life." Kelly Johnson was cited for his advancement of aeronautics.

1964   The Award of Achievement, presented by the National Aviation Club of Washington, D.C., for "outstanding achievement in airplane design and development over many years, including such models as the Constellation, P-80, F-104, JetStar, the U-2, and climaxed by the metallurgical and performance breakthroughs of the A-11 (YF-12A).

1964   The Collier Trophy (his second), following work on the 2,000-mph YF-12A interceptor. His achievement, for the previous year was called the greatest in American aviation.

1964   The Theodore von Kármán Award (his second), presented by the Air Force Association for his work with the A-11 (YF-12A) interceptor.

1964   Honorary degree of doctor of engineering, University of Michigan.

1964   Honorary degree of doctor of science, University of Southern California.

1964   Honorary degree of doctor of laws, University of California at Los Angeles.

1965   San Fernando Valley Engineer of the Year, so designated by the San Fernando, California, Valley Engineers' Council.

1965   Elected a member of the National Academy of Engineering.

1965   Elected a member of the National Academy of Sciences.

1966   The Sylvanus Albert Reed Award (his second), given by the American Institute of Aeronautics and Astronautics "in recognition of notable contributions to the aerospace sciences resulting from experimental or theoretical investigations."

1966   National Medal of Science, presented by President Lyndon B. Johnson at the White House.

1966   The Thomas D. White National Defense Award, presented by the U.S. Air Force Academy in Colorado Springs, Colo.

1967   Elected Honorary Fellow of American Institute of Aeronautics and Astronautics.

1968   Elected a Fellow of the Royal Aeronautical Society.

1969   The General William Mitchell Memorial Award, presented by the Aviators Post 743 of the American Legion at Biltmore Hotel, Wings Club, February 14.

1970   Spirit of St. Louis Medal by the American Society of Mechanical Engineers.

1970   On behalf of Lockheed's Advanced Development Projects facility, which he directed until his retirement in 1975, accepted the first annual Engineering Materials Achievement Award of the American Society for Metals. The ADP program "took titanium out of the development phase into full production for aircraft application."

1970   The Engineering Merit Award presented by the Institute for the Advancement of Engineering, Beverly Hills, Calif.

1970   Honored by the Air Force Association, Washington, D.C., for his design of the P-38 Lightning.

1971   Sixth Annual Founders Medal by the National Academy of Engineering (NAE) at the Statler-Hilton Hotel, Washington, D.C., in recognition of his fundamental contributions to engineering.

1972   The Silver Knight Award by the Lockheed Management Club of California at the Hollywood Palladium for his contributions to Lockheed's success.

1973   The first "Clarence L. Johnson Award" by the Society of Flight Test engineers in Las Vegas, Nevada, for his contributions to aviation and flight-test engineering.

1973   Civilian Kitty Hawk Memorial Award by Los Angeles Area Chamber of Commerce for his outstanding contributions in the field of aviation.

1974   Air Force Exceptional Service Award for his many outstanding contributions to the United States Air Force, from 1933 to 1974. Presented by Secretary of Air Force John McLucas.

1974   Enshrined in the Aviation Hall of Fame in Dayton, Ohio, for his outstanding contributions to aviation.

1975   The Central Intelligence Agency's Distinguished Intelligence Medal for his work on reconnaissance systems, rarely awarded outside intelligence "club."

1975   The Wright Brothers Memorial Trophy for his vital and enduring contributions over a period of 40 years to the design and development of military and commercial aircraft.

1978   The American Institute of Aeronautics and Astronautics presented "A Salute to Kelly Johnson" night—an hour-long, multi-media presentation of his career highlights.

1980   Bernt Balchen trophy, the highest award of the New York State Air Force Association, presented annually to "an individual of national prominence whose contribution to the field of aviation has been unique, extensive or of great significance." It followed announcement of the SR-71.

1981   The Department of Defense Medal for Distinguished Public Service, presented by Defense Secretary Harold Brown.

1981   Honored by the Society of Automotive Engineers (SAE) through election to the Fellow grade of membership for "his abilities to motivate a small staff to work within a tight time frame and budget in creating revolutionary aircraft designs."

1981   USAF creates "Kelly Johnson Blackbird Achievement Trophy" to "recognize the individual or group who has made the most significant contribution to the U-2, SR-71, or TR-1 program since the previous annual reunion."

1981   Daniel Guggenheim Medal "for his brilliant design of a wide range of pace-setting, commercial, combat, and reconnaissance aircraft, and for his innovative management techniques which developed these aircraft in record time at minimum cost."

1982   Meritorious Service to Aviation award from National Business Aircraft Association, recognizing design of more than 40 aircraft, including the world's first business jet, the JetStar.

1983   The Aero Club of Southern California presented the Howard Hughes Memorial Award for 1982 to Mr. C. L. "Kelly" Johnson as a leader in aviation. The recipient must have devoted a major portion of his life to the pursuit of aviation as a science and as an art. Engraved on the medal, "His vision formed the concept, his courage forged the reality."

1983   The National Security Medal was presented by President Ronald Reagan to Clarence L. Johnson for "outstanding contribution to the national intelligence effort."

1984   Honorary Royal Designer for Industry (HonRDI), in recognition of achievements in aircraft design, conferred by The Royal Society for the encouragement of Arts, Manufactures (sic) and Commerce, London.